Advances in Functional Inorganic Materials Prepared by Wet Chemical Methods

Advances in Functional Inorganic Materials Prepared by Wet Chemical Methods

Editors

Aleksej Zarkov
Aivaras Kareiva
Loreta Tamasauskaite-Tamasiunaite

MDPI • Basel • Beijing • Wuhan • Barcelona • Belgrade • Manchester • Tokyo • Cluj • Tianjin

Editors
Aleksej Zarkov
Institute of Chemistry,
Faculty of Chemistry
and Geosciences,
Vilnius University,
Vilnius, Lithuania

Aivaras Kareiva
Institute of Chemistry,
Vilnius University,
Vilnius, Lithuania

Loreta Tamasauskaite-Tamasiunaite
Department of Catalysis,
Center for Physical Sciences
and Technology,
Vilnius, Lithuania

Editorial Office
MDPI
St. Alban-Anlage 66
4052 Basel, Switzerland

This is a reprint of articles from the Special Issue published online in the open access journal *Crystals* (ISSN 2073-4352) (available at: https://www.mdpi.com/journal/crystals/special_issues/Functional_Inorganic_Materials).

For citation purposes, cite each article independently as indicated on the article page online and as indicated below:

LastName, A.A.; LastName, B.B.; LastName, C.C. Article Title. *Journal Name* **Year**, *Volume Number*, Page Range.

ISBN 978-3-0365-5623-9 (Hbk)
ISBN 978-3-0365-5624-6 (PDF)

© 2022 by the authors. Articles in this book are Open Access and distributed under the Creative Commons Attribution (CC BY) license, which allows users to download, copy and build upon published articles, as long as the author and publisher are properly credited, which ensures maximum dissemination and a wider impact of our publications.

The book as a whole is distributed by MDPI under the terms and conditions of the Creative Commons license CC BY-NC-ND.

Contents

About the Editors vii

Aleksej Zarkov, Aivaras Kareiva and Loreta Tamasauskaite-Tamasiunaite
Advances in Functional Inorganic Materials Prepared by Wet Chemical Methods
Reprinted from: *Crystals* 2021, 11, 943, doi:10.3390/cryst11080943 1

Egle Grazenaite, Edita Garskaite, Zivile Stankeviciute, Eva Raudonyte-Svirbutaviciene, Aleksej Zarkov and Aivaras Kareiva
Ga-Substituted Cobalt-Chromium Spinels as Ceramic Pigments Produced by Sol–Gel Synthesis
Reprinted from: *Crystals* 2020, 10, 1078, doi:10.3390/cryst10121078 5

Dovydas Karoblis, Ramunas Diliautas, Eva Raudonyte-Svirbutaviciene, Kestutis Mazeika, Dalis Baltrunas, Aldona Beganskiene, Aleksej Zarkov and Aivaras Kareiva
The Synthesis and Characterization of Sol-Gel-Derived SrTiO$_3$-BiMnO$_3$ Solid Solutions
Reprinted from: *Crystals* 2020, 10, 1125, doi:10.3390/cryst10121125 15

Zhi Zeng, Dongbo Wang, Jinzhong Wang, Shujie Jiao, Donghao Liu, Bingke Zhang, Chenchen Zhao, Yangyang Liu, Yaxin Liu, Zhikun Xu, Xuan Fang and Liancheng Zhao
Broadband Detection Based on 2D Bi$_2$Se$_3$/ZnO Nanowire Heterojunction
Reprinted from: *Crystals* 2021, 11, 169, doi:10.3390/cryst11020169 27

Ta Anh Tuan, Elena V. Guseva, Nguyen Anh Tien, Ho Tan Dat and Bui Xuan Vuong
Simple and Acid-Free Hydrothermal Synthesis of Bioactive Glass 58SiO$_2$-33CaO-9P$_2$O$_5$ (wt%)
Reprinted from: *Crystals* 2021, 11, 283, doi:10.3390/cryst11030283 37

Naheed Zafar, Bushra Uzair, Muhammad Bilal Khan Niazi, Ghufrana Samin, Asma Bano, Nazia Jamil, Waqar-Un-Nisa, Shamaila Sajjad and Farid Menaa
Synthesis and Characterization of Potent and Safe Ciprofloxacin-Loaded Ag/TiO$_2$/CS Nanohybrid against Mastitis Causing *E. coli*
Reprinted from: *Crystals* 2021, 11, 319, doi:10.3390/cryst11030319 49

Vera Serga, Regina Burve, Aija Krumina, Marina Romanova, Eugene A. Kotomin and Anatoli I. Popov
Extraction–Pyrolytic Method for TiO$_2$ Polymorphs Production
Reprinted from: *Crystals* 2021, 11, 431, doi:10.3390/cryst11040431 71

Constantin Buyer, David Enseling, Thomas Jüstel and Thomas Schleid
Hydrothermal Synthesis, Crystal Structure, and Spectroscopic Properties of Pure and Eu^{3+}-Doped NaY[SO$_4$]$_2$ · H$_2$O and Its Anhydrate NaY[SO$_4$]$_2$
Reprinted from: *Crystals* 2021, 11, 575, doi:10.3390/cryst11060575 85

Greta Inkrataite, Gerardas Laurinavicius, David Enseling, Aleksej Zarkov, Thomas Jüstel and Ramunas Skaudzius
Characterization of GAGG Doped with Extremely Low Levels of Chromium and Exhibiting Exceptional Intensity of Emission in NIR Region
Reprinted from: *Crystals* 2021, 11, 673, doi:10.3390/cryst11060673 107

Jolita Jablonskiene, Dijana Simkunaite, Jurate Vaiciuniene, Giedrius Stalnionis, Audrius Drabavicius, Vitalija Jasulaitiene, Vidas Pakstas, Loreta Tamasauskaite-Tamasiunaite and Eugenijus Norkus
Synthesis of Carbon-Supported MnO$_2$ Nanocomposites for Supercapacitors Application
Reprinted from: *Crystals* 2021, 11, 784, doi:10.3390/cryst11070784 117

Tamara Tsebriienko and Anatoli I. Popov
Effect of Poly(Titanium Oxide) on the Viscoelastic and Thermophysical Properties of Interpenetrating Polymer Networks
Reprinted from: *Crystals* **2021**, *11*, 794, doi:10.3390/cryst11070794 . . . . **131**

About the Editors

Aleksej Zarkov

Dr. Aleksej Zarkov is a Chief Researcher at the Institute of Chemistry of Vilnius University, Lithuania. He obtained a PhD in 2016 at Vilnius University, where he conducted research on the development of synthetic approaches for the preparation of oxide-based solid electrolytes. His research interests encompass synthesis and characterization of functional inorganic materials in the form of bulk materials, nanoparticles and thin films. Recently, A. Zarkov has focused on the synthesis and investigation of calcium phosphates for medical, optical and environmental applications. He is a co-author of 75 articles and numerous conference announcements and presentations. During his scientific career, A. Zarkov held long-term internships at various prestigious scientific institutions, including Osaka University (Osaka, Japan), Georgetown University (Washington D.C., USA), University of Cologne (Cologne, Germany), University of Aveiro (Aveiro, Portugal) and others. For his scientific achievements, A. Zarkov was awarded with several awards and scholarships from Vilnius University Rector and the Lithuanian Academy of Sciences.

Aivaras Kareiva

Prof. Aivaras Kareiva was promoted to the position of Full Professor at the Department of Inorganic Chemistry of Vilnius University in 1998. He is an expert in the sol–gel synthesis of different oxide materials. Since then, the "sol–gel chemistry" group has "prepared" 40 PhDs and 6 Postdocs who have used the sol–gel method to synthesize superconductors, optical and magnetic materials, bioceramic materials, and nanostructured materials. The sol–gel method has also been used successfully to preserve cultural heritage. He cooperated for many years with scientists from Stockholm University, the University of Tuebingen, Masaryk University Brno, Muenster University of Applied Sciences, the University of Cologne, Kyushu University, the University of Venice, the University of Strasbourg, the National Taipei University of Technology, and others. Together with co-authors, he has published more than 350 scientific articles and has participated in many research projects, grants, and contracts. In addition to scientific work, he also had some administrative positions. He was elected for the Dean position of Faculty of Chemistry (2006–2016) at Vilnius University, and from 2017 to 2022, he was also Director of Institute of Chemistry, and in 2022, he was elected as Dean of Faculty of Chemistry and Geosciences at Vilnius University. He is a Member of the Lithuanian Academy of Sciences, received five Vilnius University Rector's Research Awards (2003, 2007, 2011, 2016, 2020), two Lithuanian Republic Research Awards (2004, 2020), Lithuanian State Last Degree Fellowship (2007–2008), and Lithuanian Academy of Sciences' Juozas Matulis Award (2009). He is laureate of the Medal SAPIENTI SAT of the Association and Chapter of A. Sniadecki, K. Olszewski and Z. Wrublewski (Poland) (2016) and was awarded the Medal of the Lithuanian Academy of Sciences (2021) and Theodor von Grotthuss Medal.

Loreta Tamasauskaite-Tamasiunaite

Dr. Loreta Tamašauskaitė-Tamašiūnaitė. is currently working as a chief research associate at the Department of Catalysis at the Center for Physical Sciences and Technology (FTMC) in Vilnius (Lithuania). She has interests ranging from physical chemistry, catalysis, and electrochemistry to developing catalysts for low-temperature polymer membrane fuel cells, water splitting, and hydrogen generation from aqueous hydride solutions. She has published 124 scientific articles and is a co-author of 19 patents (https://orcid.org/0000-0001-7555-4399).

Editorial

Advances in Functional Inorganic Materials Prepared by Wet Chemical Methods

Aleksej Zarkov [1,*], Aivaras Kareiva [1] and Loreta Tamasauskaite-Tamasiunaite [2]

1. Institute of Chemistry, Vilnius University, Naugarduko 24, LT-03225 Vilnius, Lithuania; aivaras.kareiva@chgf.vu.lt
2. Center for Physical Sciences and Technology, Sauletekio Ave. 3, LT-10257 Vilnius, Lithuania; loreta.tamasauskaite@ftmc.lt
* Correspondence: aleksej.zarkov@chf.vu.lt

Citation: Zarkov, A.; Kareiva, A.; Tamasauskaite-Tamasiunaite, L. Advances in Functional Inorganic Materials Prepared by Wet Chemical Methods. *Crystals* **2021**, *11*, 943. https://doi.org/10.3390/cryst11080943

Received: 4 August 2021
Accepted: 12 August 2021
Published: 13 August 2021

Publisher's Note: MDPI stays neutral with regard to jurisdictional claims in published maps and institutional affiliations.

Copyright: © 2021 by the authors. Licensee MDPI, Basel, Switzerland. This article is an open access article distributed under the terms and conditions of the Creative Commons Attribution (CC BY) license (https://creativecommons.org/licenses/by/4.0/).

Functional inorganic materials are an indispensable part of innovative technologies, which are essential to the development of many fields of industry. The use of new materials, nanostructures, or multicomponent composites with specific chemical or physical properties promotes technological progress in electronics, optoelectronics, catalysis, biomedicine, and many other areas that are concerned with plenty of aspects of human life. Due to the broad and diverse range of the potential applications of functional inorganic materials, the development of superior synthesis pathways, reliable characterization, and a deep understanding of the structure–property relationships in materials, are rightfully considered to be fundamentally important scientific issues. Only synergetic efforts of scientists dealing with the synthesis, functionalization, and characterization of materials will lead to the development of future technologies. The Special Issue on "Advances in Functional Inorganic Materials Prepared by Wet Chemical Methods" covers a broad range of preparation routes, characterization, and the application of functional inorganic materials, as well as hybrid materials that are important in the fields of electronics, optics, biomedicine, and others.

The sol-gel method is a simple, time- and cost-effective synthetic approach providing high homogeneity and stoichiometry control of the products. These reasons make it highly suitable for the preparation of mixed-metal oxide materials and different solid solutions. For instance, Grazenaite et al. [1] employed an aqueous sol-gel method for the preparation of Ga-substituted cobalt–chromium spinels as ceramic pigments. The ion substitution resulted in significant color tuning. This demonstrated that the full substitution of Cr^{3+} by Ga^{3+} ions led to the formation of light blue powders, which yielded a violet blue color for the corresponding ceramic glaze. Karoblis et al. [2] utilized the sol-gel method for the synthesis of $(1-x)SrTiO_3$-$xBiMnO_3$ solid solutions in order to obtain new multiferroic material. The results indicated that single-phase perovskites with a cubic structure can only be synthesized up to x = 0.3. A higher $BiMnO_3$ content led to the formation of a negligible amount of the neighboring Mn_3O_4 phase. Nevertheless, some compositional trends were observed in this range. The grain size increased drastically with an increase of $BiMnO_3$, moreover, the gradual increase of the $BiMnO_3$ content resulted in noticeably higher magnetization values. Finally, using the sol-gel method, Inkrataite et al. [3] prepared cerium and chromium co-doped gadolinium–aluminum–gallium garnet (GAGG, $Gd_3Al_2Ga_3O_{12}$). The remarkable feature of this study is that an exceptionally intense emission in the near-infrared region (NIR) was achieved with an extremely low doping level. The chromium content in the obtained materials reached only 15 ppm. Another example of optical materials was provided by Buyer et al. [4]. The authors were able to synthesize Eu^{3+}-substituted $NaY[SO_4]_2$ H_2O under hydrothermal conditions. The anhydrated version of this material ($NaY[SO_4]_2$) was obtained by the post-annealing of as-prepared species. The structural and thermal properties of the synthesized compound were investigated in detail. Both compounds exhibited a strong emission in red region.

Wet chemical methods allow not only crystalline to synthesize, but also amorphous materials, which can be very attractive for medical applications. A hydrothermal approach was employed by Anh Tuan et al. [5] for the preparation of bioactive glass $58SiO_2$-$33CaO$-$9P_2O_5$. The proposed synthetic approach avoided the use of harmful acid catalysts and was confirmed as one of the ideal methods for the preparation of ternary bioactive glass. For specific medical applications more complex hybrid organic–inorganic structures can be used. For example, the successful preparation of the biocompatible nanohybrid of ciprofloxacin-Ag/TiO_2/chitosan was demonstrated by Zafar et al. [6]. The obtained nanaheterostructures were shown to be effective against mastitis causing *E. coli*.

Titanium dioxide, besides being used in heterostructures for biomedical applications, is a very technologically important material, which can be utilized for a broad range of applications. In this light, simple and cost-effective methods for the synthesis of TiO_2 are highly desirable. Serga et al. [7] demonstrated for the first time the possibilities of the extraction–pyrolytic method (EPM) for the production of nanocrystalline TiO_2 powders. It was observed that the EPM permitted the production of both monophase (anatase or rutile polymorph) and biphase (mixed anatase–rutile polymorphs) nanocrystalline TiO_2. The influence of poly(titanium oxide) obtained using the sol-gel method on the viscoelastic and thermophysical properties of interpenetrating polymer networks (IPNs) based on crosslinked polyurethane (PU) and poly(hydroxyethyl methacrylate) (PHEMA), was studied by Tsebriienko and Popov [8]. It was found that an increase in poly(titanium oxide) content led to a decrease in the intensity of the relaxation maximum for PHEMA phase and an increase in the effective crosslinking density due to the partial grafting of the inorganic component to acrylate.

Carbon-supported MnO_2 nanocomposites for the application of supercapacitors were fabricated by Jablonskiene et al. [9] using the microwave-assisted heating method. For comparison, the nanocomposites were synthesized by one-step and two-step approaches. The high specific capacitance of 980.7 F g^{-1} was achieved from cyclic voltammetry measurements, whereas the specific capacitance of 949.3 F g^{-1} at 1 A g^{-1} was obtained from a galvanostatic charge/discharge test. The specific capacitance retention was 93% after 100 cycles at 20 A g^{-1}, indicating good electrochemical stability.

Zeng et al. [10] utilized hydrothermally synthesized ZnO nanowires for the fabrication of 2D Bi_2Se_3/ZnO heterojunction, which was employed for the broadband photodetection. The fabricated heterojunction device demonstrated not only an enhanced photoresponsivity of 0.15 A/W at 377 nm, which was three times higher than that of bare ZnO nanowire (0.046 A/W), but also achieved a broadband photoresponse from UV to near-infrared region was achieved.

The present Special Issue on "Advances in Functional Inorganic Materials Prepared by Wet Chemical Methods" demonstrates the versatility of wet chemical methods for the preparation of functional materials for a broad range of applications. It can be considered as a status report reviewing the progress that has been achieved in the processing of inorganic materials.

Author Contributions: Conceptualization, A.Z., A.K. and L.T.-T.; writing—original draft preparation, A.Z., A.K. and L.T.-T.; writing—review and editing, A.Z., A.K. and L.T.-T.; funding acquisition, A.Z. and A.K. All authors have read and agreed to the published version of the manuscript.

Funding: This project has received funding from European Social Fund (project No. 09.3.3-LMT-K-712-19-0069) under grant agreement with the Research Council of Lithuania (LMTLT).

Conflicts of Interest: The authors declare no conflict of interest.

References

1. Grazenaite, E.; Garskaite, E.; Stankeviciute, Z.; Raudonyte-Svirbutaviciene, E.; Zarkov, A.; Kareiva, A. Ga-substituted cobalt-chromium spinels as ceramic pigments produced by sol-gel synthesis. *Crystals* **2020**, *10*, 1078. [CrossRef]
2. Karoblis, D.; Diliautas, R.; Raudonyte-Svirbutaviciene, E.; Mazeika, K.; Baltrunas, D.; Beganskiene, A.; Zarkov, A.; Kareiva, A. The Synthesis and characterization of sol-gel-derived $SrTiO_3$-$BiMnO_3$ solid solutions. *Crystals* **2020**, *10*, 1125. [CrossRef]

3. Inkrataite, G.; Laurinavicius, G.; Enseling, D.; Zarkov, A.; Jüstel, T.; Skaudzius, R. Characterization of GAGG doped with extremely low levels of chromium and exhibiting exceptional intensity of emission in NIR region. *Crystals* **2021**, *11*, 673. [CrossRef]
4. Buyer, C.; Enseling, D.; Jüstel, T.; Schleid, T. Hydrothermal synthesis, crystal structure, and spectroscopic properties of pure and Eu^{3+}-doped NaY$[SO_4]_2$ H2O and Its Anhydrate NaY$[SO_4]_2$. *Crystals* **2021**, *11*, 575. [CrossRef]
5. Anh Tuan, T.; Guseva, E.V.; Anh Tien, N.; Tan Dat, H.; Vuong, B.X. Simple and acid-free hydrothermal synthesis of bioactive glass 58SiO_2-33CaO-9P_2O_5 (wt%). *Crystals* **2021**, *11*, 283. [CrossRef]
6. Zafar, N.; Uzair, B.; Niazi, M.B.K.; Samin, G.; Bano, A.; Jamil, N.; Un-Nisa, W.; Sajjad, S.; Menaa, F. Synthesis and characterization of potent and safe ciprofloxacin-loaded Ag/TiO_2/CS nanohybrid against mastitis causing *E. coli*. *Crystals* **2021**, *11*, 319. [CrossRef]
7. Serga, V.; Burve, R.; Krumina, A.; Romanova, M.; Kotomin, E.A.; Popov, A.I. Extraction–pyrolytic method for TiO_2 polymorphs production. *Crystals* **2021**, *11*, 431. [CrossRef]
8. Tsebriienko, T.; Popov, A.I. Effect of poly(titanium oxide) on the viscoelastic and thermophysical properties of interpenetrating polymer networks. *Crystals* **2021**, *11*, 794. [CrossRef]
9. Jablonskiene, J.; Simkunaite, D.; Vaiciuniene, J.; Stalnionis, G.; Drabavicius, A.; Jasulaitiene, V.; Pakstas, V.; Tamasauskaite-Tamasiunaite, L.; Norkus, E. Synthesis of carbon-supported mno_2 nanocomposites for supercapacitors application. *Crystals* **2021**, *11*, 784. [CrossRef]
10. Zeng, Z.; Wang, D.; Wang, J.; Jiao, S.; Liu, D.; Zhang, B.; Zhao, C.; Liu, Y.; Liu, Y.; Xu, Z.; et al. Broadband detection based on 2D Bi_2Se_3/ZnO nanowire heterojunction. *Crystals* **2021**, *11*, 169. [CrossRef]

Article

Ga-Substituted Cobalt-Chromium Spinels as Ceramic Pigments Produced by Sol–Gel Synthesis

Egle Grazenaite [1,2], Edita Garskaite [3], Zivile Stankeviciute [2], Eva Raudonyte-Svirbutaviciene [4], Aleksej Zarkov [2] and Aivaras Kareiva [2,*]

[1] Department of Detailed Research, Cultural Heritage Centre, Asmenos 10, 01135 Vilnius, Lithuania; eglolis@gmail.com
[2] Institute of Chemistry, Vilnius University, Naugarduko 24, 03225 Vilnius, Lithuania; zivile.stankeviciute@chf.vu.lt (Z.S.); aleksej.zarkov@chf.vu.lt (A.Z.)
[3] Wood Science and Engineering, Department of Engineering Sciences and Mathematics, Luleå University of Technology, Forskargatan 1, 931 87 Skellefteå, Sweden; edita.garskaite@ltu.se
[4] SRI Nature Research Centre, Institute of Geology and Geography, Akademijos 2, 08412 Vilnius, Lithuania; eva.raudonyte@gmail.com
* Correspondence: aivaras.kareiva@chgf.vu.lt

Received: 27 October 2020; Accepted: 24 November 2020; Published: 25 November 2020

Abstract: For the first time to the best of our knowledge, cobalt-chromium spinels $CoCr_{2-x}Ga_xO_4$ with different amounts of gallium (x = 0–2 with a step of 0.5) were synthesized via the aqueous sol–gel route as ceramic pigments. The phase composition, crystallite size, morphological features, and color parameters of new compositions and their corresponding ceramic glazes were investigated using XRD, CIELab, SEM, and optical microscopy. It was demonstrated that the formation of single-phase $CoCr_{2-x}Ga_xO_4$ samples was problematic. Full substitution of Cr^{3+} by Ga^{3+} ion in the spinel resulted in the formation of light blue powders, which yielded violetish blue color for the corresponding ceramic glaze.

Keywords: sol–gel processing; cobalt chromite; mixed-metal oxides; gallium substitution; ceramic pigments

1. Introduction

Different metal oxides and mixed metal oxides are known to serve as ceramic pigments. For various applications, pigments have specific requirements such as chemical and thermal stability, particle size, hiding and tinting power, etc. Spinels which are mixed-metal oxides with a general formula of AB_2O_4 are very attractive in the pigmentary field due to their characteristics of high mechanical resistance, and high thermal and chemical stability [1,2]. The nature of tetrahedral or octahedral cations in spinel structure and the potential of different types of doping give diversity in colors and properties. Cobalt chromite ($CoCr_2O_4$) pigments are well known and have been synthesized using sol-gel [3–5], combustion [6–8], combined sol-gel combustion [9], solid-state reaction [10], microwave-assisted [11], and spray pyrolysis [12] methods. In our previous studies, cobalt chromite based compounds $Co_{1-x}M_xCr_2O_4$ (M = Ni, Cu, and Zn) with different transition metal concentrations (0 ≤ x ≤ 1 with a step of 0.25) [13] and $CoCr_{2-x}Ln_xO_4$ (Ln = Tm^{3+} and Yb^{3+}) pigments with different substitutional levels of lanthanide (x = 0–0.5) [14] have been synthesized using an aqueous sol–gel synthetic approach and characterized by various techniques.

Gallium-containing spinels attracted the interest of scientists for many decades. Such compounds were synthesized and investigated mainly for conducting luminescence and other properties. $CdGa_2O_4$ spinel was found to be a promising compound as a transparent electronic conductor [15], whereas $ZnGa_2O_4$ is a great UV-transparent electronic conductor [16]. $MgGa_2O_4$ doped by

Cr^{3+} ion [17] and Si^{4+} ion co-doped $MgGa_2O_4$:Cr^{3+} [18] were prepared and investigated as phosphors. $CuGa_2O_4$ nanocrystalline powders were synthesized and investigated as sensors for H_2, liquefied petroleum gas, and NH_3 [19]. Moreover, $NiGa_2O_4$ thin films doped with different levels of Eu^{3+} ion were prepared and their luminescent properties were characterized [20]. However, there are no records of the research preparing gallium-containing spinels as ceramic pigments, to the best of our knowledge. In general, the number of reports of the research of any gallium-containing structures as ceramic pigments is relatively low. Lutetium gallium garnets co-doped with chromium and calcium ($Ca_xCr_xLu_{3-2x}Ga_5O_{12}$ up to x = 0.2) were obtained by solid-state reaction as pink ceramic pigments [21,22]. In another study, the investigated gallium gadolinium garnet ($Gd_3Ga_5O_{12}$) doped with Cr^{4+} resulted in green shades of ceramic glazes due to reduction of Cr^{4+} to Cr^{3+} that dissolved in the glaze [23]. Perovskite-like purple inorganic pigments $YGa_{1-x}Mn_xO_3$ (0 < x ≤ 0.10) were prepared by a sol–gel technique [24].

In this study, Ga^{3+} was chosen as a substitutional ion for the modification of cobalt chromite by replacing Cr^{3+} ion. Therefore, cobalt-chromium spinels as ceramic pigments $CoCr_{2-x}Ga_xO_4$ with different substitutional levels of gallium (x = 0–2 with a step of 0.5) were synthesized using the aqueous sol–gel method. The phase purity, morphological properties, and color parameters of new $CoCr_{2-x}Ga_xO_4$ ceramic pigments were investigated in this study.

2. Materials and Methods

2.1. Materials

All purchased reagents were used as received without further purification. Aqueous sol–gel synthesis [16] was carried out using $Cr(NO_3)_3 \cdot 9H_2O$ (99.0%, Sigma-Aldrich, Darmstadt, Germany), $Co(NO_3)_2 \cdot 6H_2O$ (97.7%, Alfa Aesar, Kandel, Germany), Ga_2O_3 (99.99%, Alfa Aesar, Kandel, Germany), HNO_3 (67%, Eurochemicals, Vilnius, Lithuania) and 1,2-ethanediol $C_2H_6O_2$ (99.5%, Sigma-Aldrich, Darmstadt, Germany) as starting materials for the preparation of precursor gels. For the formation of ceramic glazes, the Czech transparent colorless base glaze (Ferro, Frankfurt/Main, Germany) was used.

2.2. Aqueous Sol–Gel Synthesis

For the synthesis of Co-Cr-O precursor gel, stoichiometric amounts of $Co(NO_3)_2 \cdot 6H_2O$ and $Cr(NO_3)_3 \cdot 9H_2O$ were dissolved in deionized water and mixed together [4]. For the synthesis of Co-Cr-Ga-O precursor gel, the appropriate amount of Ga_2O_3 was dissolved in diluted hot nitric acid first and then mixed with aqueous solutions of $Co(NO_3)_2 \cdot 6H_2O$ and $Cr(NO_3)_3 \cdot 9H_2O$. After mixing, the solutions were stirred at 40–50 °C for 20 min and then 2 mL of 1,2-ethanediol was added with continuous stirring at the same temperature for 1 h. The solutions were concentrated by continuous stirring and evaporation at 60–70 °C. Prepared gels were dried in a furnace at 105–110 °C in air, carefully ground in an agate mortar, and annealed at 700 °C in the air for 3 h with a heating rate of 5 °C/min. The obtained powders were ground once again and additionally heated at 1000 °C in the air for 5 h with a heating rate of 10 °C/min.

2.3. Preparation of Ceramic Glazes

The obtained pigments were used for the preparation of ceramic glazes. For that purpose, 0.05 g (5 wt%) of each pigment was mixed with 0.95 g of the Czech base glaze powders and a little bit of water and carefully plastered onto terracotta tiles (0.03 × 0.04 m). After drying in air, the prepared terracotta samples were fired in an oxidizing atmosphere in an electric furnace at 1000 °C for 1 h with a heating rate of 5 °C/min.

2.4. Characterization

For the identification of the phase composition of the resulted products, the powder X-ray diffraction (XRD) analysis was used. The measurements were performed using a Rigaku MiniFlex II diffractometer (The Woodlands, TX, USA), operated at 30 kV and 10 mA with a scanning speed of 10 °/min, in a scanning range of 2θ = 10–80°, using Cu Kα radiation (λ = 1.540562 Å). The obtained diffraction data were refined by the Rietveld method using the FullProf suite. The tentative crystallite sizes were determined by the Scherrer equation:

$$\tau = 0.9\lambda/B\cos\theta \qquad (1)$$

where τ is the mean crystallite size, λ is the X-ray wavelength, B is the line broadening at half maximum intensity (FWHM) (in radians) and θ is the Bragg angle. The color of the pigments and the ceramic glazes was evaluated by the CIELab colorimetric method, which is recommended by the Commission Internationale de l'Eclairage. The L*, a*, and b* parameters were measured on a Perkin Elmer Lambda 950 spectrophotometer (Waltham, MA, USA) in the 780–380 nm range, employing an illuminant D65 and a 10° standard observer. In the CIELab system, the coordinate L* represents the lightness of the color (L* = 0 and L* = 100 represents black and white, respectively). The negative/positive values of coordinate a* represent green/red hue, respectively, and the parameter b* corresponds to blue/yellow hue, where negative values are for blue and positive for yellow. The morphological features of obtained samples were investigated using a scanning electron microscope (SEM) (Hitach SU70, Tokyo, Japan). Quantification of Co, Cr, and Ga in synthesized specimens was performed by inductively coupled plasma optical emission spectrometry (ICP-OES) using Perkin-Elmer Optima 7000 DV spectrometer (Waltham, MA, USA). Sample decomposition procedure was carried out in concentrated nitric acid (HNO_3, Rotipuran® Supra 69%, Roth) using microwave reaction system Anton Paar Multiwave 3000 (Graz, Austria) equipped with XF100 rotor and PTFE liners. The following program was used for the dissolution of powders: during the first step, microwave power was linearly increased to 800 W in 15 min and held at this point for the next 20 min. Once the vessels have been fully cooled and depressurized the obtained clear solutions were quantitatively transferred into volumetric flasks of a certain volume and diluted with deionized water. Calibration solutions were prepared by an appropriate dilution of the stock standard solutions (single-element ICP standards 1000 mg/L, Roth).

3. Results and Discussion

The XRD patterns of Ga-doped $CoCr_{2-x}Ga_xO_4$ (x = 0–2 with a step of 0.5) samples, depending on the substitution ratio and heating temperature, are given in Figure 1. The main crystalline phase of the synthesis products obtained at 700 °C was a solid solution of cubic $CoCr_2O_4$ (PDF 22-1084) and $CoGa_2O_4$ spinels (PDF 11-0698). However, an additional Cr_2O_3 phase (PDF 38-1479) was observed for the sample with x = 0.5. Phase composition analysis revealed that chromium substitution by gallium was not successful at a higher temperature. Additional Ga_2O_3 phase was formed almost at all substitutional levels. Interestingly, the XRD patterns of the samples with x = 1–2 annealed at 1000 °C showed a minor amount of Ga_2O_3 crystalline phase (PDF 41-1103) (see Figure 1b), which was not observed in the samples heated at 700 °C.

The cubic cell parameters were calculated for all Ga-doped $CoCr_{2-x}Ga_xO_4$ samples. It is interesting to note that the cell parameter a = 8.4582(3) Å determined for the $CoCr_2O_4$ was very similar to the cell parameter of the gallium-substituted samples. Thus, with increasing Ga^{3+} amount in the spinel structure, no monotonical shift of the diffraction peaks to higher or lower 2θ values were observed. This is not surprising, since the ionic radii of Cr^{3+} and Ga^{3+} ions in VI-fold coordination are almost identical (0.615 Å for Cr^{3+} and 0.620 Å for Ga^{3+}) [25]. The estimated crystallite size of the spinel phase for all compositions ranged from 11.4 to 32.6 nm and from 46.6 to 54.4 nm for the samples obtained at 700 and 1000 °C, respectively. The crystallite size decreased linearly with the increase of substitution level (see Table 1).

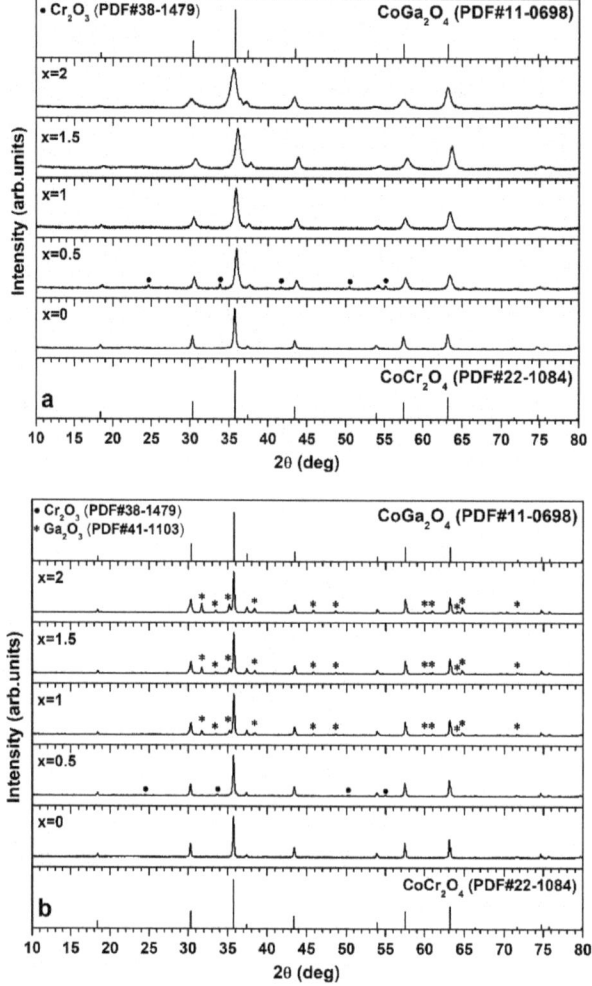

Figure 1. XRD patterns of CoCr$_{2-x}$Ga$_x$O$_4$ powders (x = 0–2) calcined at 700 °C (**a**) and annealed at 1000 °C (**b**).

Table 1. The estimated crystallite size of the CoCr$_{2-x}$Ga$_x$O$_4$ powders synthesized at different temperatures.

CoCr$_{2-x}$Ga$_x$O$_4$ Sample	Crystallite Size, nm	
	700 °C	1000 °C
x = 0	32.6 ± 0.5	54.4 ± 0.7
x = 0.5	27.9 ± 0.6	51.9 ± 0.5
x = 1.0	22.4 ± 0.4	49.5 ± 0.5
x = 1.5	16.8 ± 0.5	48.0 ± 0.6
x = 2.0	11.4 ±.0.3	46.6 ± 0.7

To confirm the chemical composition of the synthesized samples, the elemental analysis using ICP-OES was performed. The results of the analysis are summarized in Table 2. Since Co was not changed in the synthesized series, the molar ratio of the elements was normalized by the concentration of Co. It is seen that the determined ratio is very close to the nominal values for all samples, which indicates that the suggested synthesis approach is suitable for the preparation of $CoCr_{2-x}Ga_xO_4$ powders with the controllable chemical composition of the final products.

Table 2. The results of the elemental analysis of $CoCr_{2-x}Ga_xO_4$ powders by ICP-OES.

$CoCr_{2-x}Ga_xO_4$ Sample	n (Co)	n (Cr)	n (Ga)
x = 0	1	2.01	-
x = 0.5	1	1.48	0.510
x = 1.0	1	1.02	1.03
x = 1.5	1	0.491	1.52
x = 2.0	1	-	2.03

The SEM micrographs of selected pigment samples (x = 0.5 and 2) obtained at different heating temperatures are given in Figure 2. At the low heating temperature, the surface of synthesized pigments is composed of irregularly shaped particles which are highly agglomerated independently of the substitution level. The powders with the lowest substitution ratio (Figure 2a) are composed of particles with an irregular shape. The particles of the sample annealed at 1000 °C with the highest substitution ratio (Figure 2d) are needle-like plates. Nevertheless, the agglomeration of the particles is very high and no separate particles could be clearly distinguished.

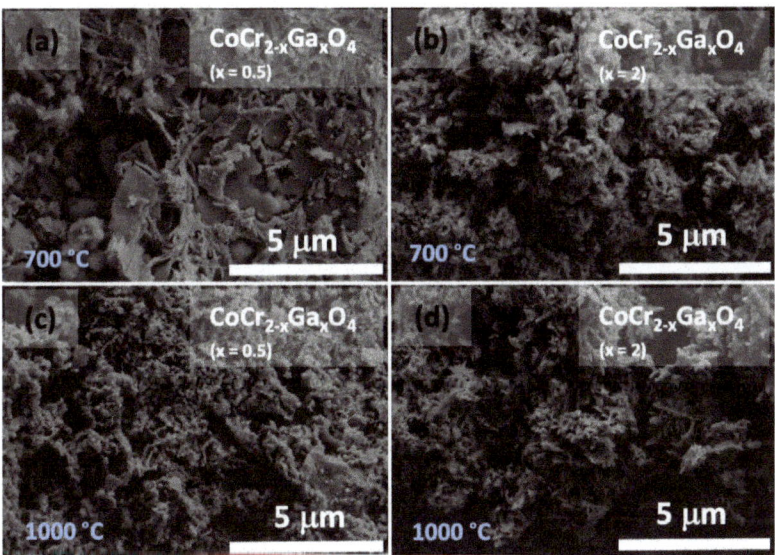

Figure 2. SEM micrographs of sol–gel derived $CoCr_{2-x}Ga_xO_4$ pigments, when x = 0.5 (a,c) and x = 2 (b,d), obtained at 700 °C (a,b) and 1000 °C (c,d).

The representative SEM micrograph and a digital picture of optical microscopy of the ceramic glaze obtained using sol–gel derived fully substituted $CoGa_2O_4$ pigment are presented in Figure 3. SEM investigation revealed good dispersion of the pigments over ceramic tiles. However, the separate particles are visible in the $CoGa_2O_4$ glaze (Figure 3a). Optical microscopy confirmed the presence of pigment particles on the surface and negligible formation of gas bubbles. No cracks, caves, or other physical defects could be observed on the surface of ceramic glazes, notwithstanding the mentioned bubbles and separate particles.

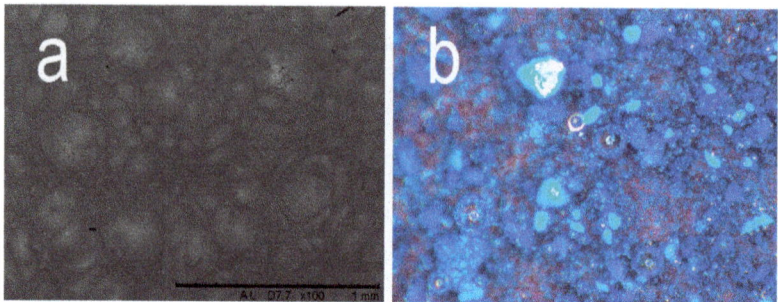

Figure 3. Images of the glaze prepared using $CoGa_2O_4$ obtained by SEM (X 100) (**a**) and optical microscopy (X 60) (**b**).

The colorimetric parameters of the pigments obtained at 1000 °C and corresponding glazes are summarised in Table 3. As it was presumed, the most divergent results of the CIELab measurements are of the pigment and glaze samples when x = 0.5. The impurity of Cr_2O_3 gives a greener hue to the mentioned specimens, reasoning the most negative values of parameter a^*, comparing to the results of other samples with x = 1–2. The general tendency of increase of lightness parameter L^* for the pigments and decrease for the glazes with an increase of substitution ratio is well observed. Moreover, for the pigments the values of parameter b^* are increasingly negative with the increase of gallium concentration, implying the enhancement of blue hue. On the contrary, for the glazes the values of parameter a^* convert into positive values, meaning the enhancement of red hue. Corresponding to the CIELab results, the pigments give diversity in colors from bluish-green to light blue (Figure 4).

Table 3. CIELab colorimetric parameters of the $CoCr_{2-x}Ga_xO_4$ pigments, obtained at 1000 °C, and corresponding glazes.

Amount of Ga (x)	Pigment			Glaze		
	L^*	a^*	b^*	L^*	a^*	b^*
0	47.69	−8.31	−3.89	33.47	−10.92	−10.36
0.5	50.25	−12.39	−7.56	30.86	−7.59	−4.41
1	48.48	−8.49	−5.60	31.53	−4.01	−4.16
1.5	52.48	−11.83	−8.36	30.74	−4.32	−6.84
2	54.44	−9.48	−14.07	27.69	3.67	−9.35

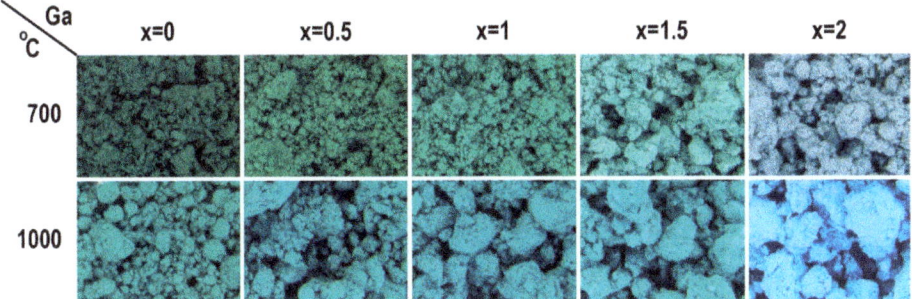

Figure 4. Digital photographs of Ga-doped cobalt chromite pigments obtained at different heating temperatures.

The pigments annealed at 1000 °C possess brighter colors. The confirmation of the CIELab results of ceramic glazes is given in Figure 5. The outstanding glazes are obtained with different substitutional levels of gallium. As was expected, the glaze with $CoCr_{1.5}Ga_{0.5}O_4$ pigment possesses a green hue due to the additional chromium(III) oxide phase. However, the most unexpected results, concerning the colors of the pigments, were with the fully substituted pigmented glaze, which turned out to be violetish blue.

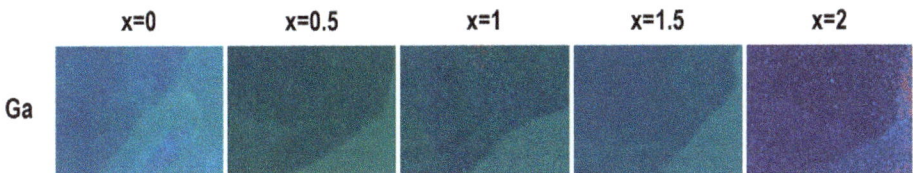

Figure 5. Digital photographs of the ceramic glazes prepared with Ga-doped pigments.

4. Conclusions

The attempts to synthesize the single-phase $CoCr_{2-x}Ga_xO_4$ samples using sol–gel method were done in this study. The main crystalline phase of the synthesis products obtained at 700 °C was a solid solution of cubic $CoCr_2O_4$ and $CoGa_2O_4$ spinels. However, an additional Cr_2O_3 phase was observed for the sample with $x = 0.5$. Phase composition analysis revealed that chromium substitution by gallium was not successful at a higher temperature either. A minor amount of Ga_2O_3 crystalline phase was detected in XRD patterns of the samples with $x = 1$–2 annealed at 1000 °C. SEM analysis revealed that the agglomeration of the particles is very high and no separate particles could be clearly distinguished. SEM micrographs and optical microscopy investigation revealed well dispersion of the pigments within ceramic glazes prepared using $CoCr_{2-x}Ga_xO_4$ pigments. The formation of high-quality glazes was confirmed. However, the colors of obtained $CoCr_{2-x}Ga_xO_4$ pigments and their corresponding ceramic glazes were unexpected. The substitution of Cr^{3+} by Ga^{3+} ion led to the gradual light blueness of the pigment. The impurity of Cr_2O_3 in the sample with the lowest substitution ratio gave the green hue to both the pigment and the corresponding ceramic glaze. Surprisingly, the ceramic glaze prepared using fully substituted $CoGa_2O_4$ pigment turned out to possess violetish blue color.

Author Contributions: Conceptualization, E.G. (Egle Grazenaite) and A.K.; methodology, E.G. (Egle Grazenaite) and E.G. (Edita Garskaite); formal analysis, E.G. (Egle Grazenaite), Z.S., E.R.-S., A.Z., and A.K.; investigation, E.G. (Egle Grazenaite), Z.S., E.R.-S., and A.Z.; resources, E.G. (Edita Garskaite), A.K.; writing—original draft preparation, E.G. (Egle Grazenaite); writing—review and editing, A.K.; supervision, A.K.; funding acquisition, E.G. (Edita Garskaite), A.K. All authors have read and agreed to the published version of the manuscript.

Funding: The work has been partially supported by the Swedish Research Council for Environment, Agricultural Sciences and Spatial Planning (FORMAS) Project "Utilization of solid inorganic waste from the aquaculture industry as wood reinforcement material for flame retardancy" (grant no. 2018-01198).

Acknowledgments: Authors would like to thank Gediminas Kreiza for their technical assistance and helpful discussions.

Conflicts of Interest: The authors declare no conflict of interest.

References

1. Lorenzi, G.; Baldi, G.; Di Benedetto, F.; Faso, V.; Lattanzi, P.; Romanelli, M. Spectroscopic study of a Ni-bearing gahnite pigment. *J. Eur. Ceram. Soc.* **2006**, *26*, 317–321. [CrossRef]
2. DeSouza, L.; Zamian, J.; Filho, G.N.D.R.; Soledade, L.E.; DosSantos, I.; Souza, A.; Scheller, T.; Angélica, R.S.; Dacosta, C. Blue pigments based on $Co_xZn_{1-x}Al_2O_4$ spinels synthesized by the polymeric precursor method. *Dye. Pigment.* **2009**, *81*, 187–192. [CrossRef]
3. Cui, H.; Zayat, M.; Levy, D. Sol-Gel Synthesis of Nanoscaled Spinels Using Propylene Oxide as a Gelation Agent. *J. Sol-Gel Sci. Technol.* **2005**, *35*, 175–181. [CrossRef]
4. Jasaitis, D.; Beganskiene, D.; Senvaitiene, J.; Kareiva, A.; Ramanauskas, R.; Juskenas, R.; Selskis, A. Sol-gel synthesis and characterization of cobalt chromium spinel CoCr2O4. *Chemija* **2011**, *22*, 125–130.
5. Hedayati, H.; Alvani, A.A.S.; Sameie, H.; Salimi, R.; Moosakhani, S.; Tabatabaee, F.; Zarandi, A.A. Synthesis and characterization of $Co_{1-x}Zn_xCr_{2-y}Al_yO_4$ as a near-infrared reflective color tunable nano-pigment. *Dye. Pigment.* **2015**, *113*, 588–595. [CrossRef]
6. Hu, D.-S.; Han, A.-J.; Ye, M.; Chen, H.-H.; Zhang, W. 2-xAlxO4 by Low-temperature Combustion Synthesis. *J. Inorg. Mater.* **2011**, *26*, 285–289. [CrossRef]
7. Chamyani, S.; Salehirad, A.; Oroujzadeh, N.; Fateh, D.S. Effect of fuel type on structural and physicochemical properties of solution combustion synthesized $CoCr_2O_4$ ceramic pigment nanoparticles. *Ceram. Int.* **2018**, *44*, 7754–7760. [CrossRef]
8. Miranda, E.A.C.; Sepúlveda, A.A.L.; Carvajal, J.F.M.; Gil, S.V.; Baena, O.J.R. Green inorganic pigment production with spinel structure $CoCr_2O_4$ by solution combustion synthesis. *TECCIENCIA* **2019**, *14*, 41–51. [CrossRef]
9. Lei, S.-L.; Liang, G.-J.; Wang, Y.; Zhou, S.-X.; Zhang, X.; Li, S. Sol-gel combustion synthesis and characterization of $CoCr_2O_4$ ceramic powder used as color solar absorber pigment. *Optoelectron. Lett.* **2020**, *16*, 365–368. [CrossRef]
10. Mindru, I.; Gingasu, D.; Marinescu, G.; Patron, L.; Calderon-Moreno, J.M.; Bartha, C.; Andronescu, C.; Crişan, A. Cobalt chromite obtained by thermal decomposition of oxalate coordination compounds. *Ceram. Int.* **2014**, *40*, 15249–15258. [CrossRef]
11. Tanisan, B.; Dondi, M. Cobalt chromite nano pigments synthesis through microwave-assisted polyol route. *J. Sol-Gel Sci. Technol.* **2017**, *83*, 590–595. [CrossRef]
12. Betancur-Granados, N.; Restrepo-Baena, O.J. Flame spray pyrolysis synthesis of ceramic nanopigments $CoCr_2O_4$: The effect of key variables. *J. Eur. Ceram. Soc.* **2017**, *37*, 5051–5056. [CrossRef]
13. Grazenaite, E.; Pinkas, J.; Beganskiene, A.; Kareiva, A. Sol–gel and sonochemically derived transition metal (Co, Ni, Cu, and Zn) chromites as pigments: A comparative study. *Ceram. Int.* **2016**, *42*, 9402–9412. [CrossRef]
14. Grazenaite, E.; Jasulaitiene, V.; Ramanauskas, R.; Kareiva, A. Sol–gel synthesis, characterization and application of lanthanide-doped cobalt chromites ($CoCr_{2-x}Ln_xO_4$; Ln = Tm^{3+} and Yb^{3+}). *J. Eur. Ceram. Soc.* **2018**, *38*, 3361–3368. [CrossRef]
15. Omata, T.; Ueda, N.; Hikuma, N.; Ueda, K.; Mizoguchi, H.; Hashimoto, T.; Kawazoe, H. New oxide phase with wide band gap and high electroconductivity $CdGa_2O_4$ spinel. *Appl. Phys. Lett.* **1993**, *62*, 499–500. [CrossRef]
16. Omata, T.; Ueda, N.; Ueda, K.; Kawazoe, H. New ultraviolet–transport electroconductive oxide, $ZnGa_2O_4$ spinel. *Appl. Phys. Lett.* **1994**, *64*, 1077–1078. [CrossRef]
17. Mondal, A.; Manam, J. Investigations on spectroscopic properties and temperature dependent photoluminescence of Cr^{3+} doped $MgGa_2O_4$ phosphor. *Mater. Res. Express* **2019**, *6*, 095081. [CrossRef]
18. Mondal, A.; Manam, J. Structural and Luminescent Properties of Si^{4+} Co-Doped $MgGa_2O_4$:Cr^{3+} Near Infra-Red Long Lasting Phosphor. *ECS J. Solid State Sci. Technol.* **2017**, *6*, R88–R95. [CrossRef]

19. Biswas, S.K.; Sarkar, A.; Pathak, A.; Pramanik, P. Studies on the sensing behaviour of nanocrystalline $CuGa_2O_4$ towards hydrogen, liquefied petroleum gas and ammonia. *Talanta* **2010**, *81*, 1607–1612. [CrossRef]
20. Cabello, G.; Lillo-Arroyo, L.; Caro-Díaz, C.; Valenzuela-Melgarejo, F.; Fernández-Pérez, A.; Buono-Core, G.; Chornik, B. Study on the photochemical preparation of nickel gallium oxide spinel doped with Eu(III) ions from carboxylate and β-diketonate complexes and the evaluation of its optical properties. *Thin Solid Films* **2018**, *647*, 33–39. [CrossRef]
21. Galindo, R.; Llusar, M.; Tena, M.A.; Monros, G.; Badenes, J. New pink ceramic pigment based on chromium (IV)-doped lutetium gallium garnet. *J. Eur. Ceram. Soc.* **2007**, *27*, 199–205. [CrossRef]
22. Galindo, R.; Badenes, J.; Llusar, M.; Tena, M.; Ángeles; Monros, G. Synthesis and characterisation of chromium lutetium gallium garnet solid solution. *Mater. Res. Bull.* **2007**, *42*, 437–445. [CrossRef]
23. Monrós, G.; Pinto, H.; Badenes, J.; Llusar, M.; Tena, M.; Ángeles; March, J.A.B. Chromium(IV) Stabilisation in New Ceramic Matrices by Coprecipitation Method: Application as Ceramic Pigments. *Z. Anorg. Allg. Chem.* **2005**, *631*, 2131–2135. [CrossRef]
24. Tamilarasan, S.; Sarma, D.; Reddy, M.L.P.; Natarajan, S.; Gopalakrishnan, J. $YGa_{1-x}Mn_xO_3$: A novel purple inorganic pigment. *RSC Adv.* **2013**, *3*, 3199–3202. [CrossRef]
25. Shannon, R.D. Revised effective ionic radii and systematic studies of interatomic distances in halides and chalcogenides. *Acta Crystallogr. Sect. A* **1976**, *32*, 751–767. [CrossRef]

Publisher's Note: MDPI stays neutral with regard to jurisdictional claims in published maps and institutional affiliations.

© 2020 by the authors. Licensee MDPI, Basel, Switzerland. This article is an open access article distributed under the terms and conditions of the Creative Commons Attribution (CC BY) license (http://creativecommons.org/licenses/by/4.0/).

Article

The Synthesis and Characterization of Sol-Gel-Derived SrTiO$_3$-BiMnO$_3$ Solid Solutions

Dovydas Karoblis [1], Ramunas Diliautas [1], Eva Raudonyte-Svirbutaviciene [2], Kestutis Mazeika [3], Dalis Baltrunas [3], Aldona Beganskiene [1], Aleksej Zarkov [1,*] and Aivaras Kareiva [1]

[1] Institute of Chemistry, Vilnius University, Naugarduko 24, LT-03225 Vilnius, Lithuania; dovydas.karoblis@chgf.vu.lt (D.K.); ramunas.diliautas@bioeksma.lt (R.D.); aldona.beganskiene@chf.vu.lt (A.B.); aivaras.kareiva@chgf.vu.lt (A.K.)
[2] SRI Nature Research Centre, Institute of Geology and Geography, Akademijos 2, LT-08412 Vilnius, Lithuania; eva.svirbutaviciene@gamtc.lt
[3] Center for Physical Sciences and Technology, LT-02300 Vilnius, Lithuania; kestas@ar.fi.lt (K.M.); dalis@ar.fi.lt (D.B.)
* Correspondence: aleksej.zarkov@chf.vu.lt

Received: 13 November 2020; Accepted: 9 December 2020; Published: 10 December 2020

Abstract: In this study, the aqueous sol-gel method was employed for the synthesis of $(1-x)$SrTiO$_3$-xBiMnO$_3$ solid solutions. Powder X-ray diffraction analysis confirmed the formation of single-phase perovskites with a cubic structure up to $x = 0.3$. A further increase of the BiMnO$_3$ content led to the formation of a negligible amount of neighboring Mn$_3$O$_4$ impurity, along with the major perovskite phase. Infrared (FT-IR) analysis of the synthesized specimens showed gradual spectral change associated with the superposition effect of Mn-O and Ti-O bond lengths. By introducing BiMnO$_3$ into the SrTiO$_3$ crystal structure, the size of the grains increased drastically, which was confirmed by means of scanning electron microscopy. Magnetization studies revealed that all solid solutions containing the BiMnO$_3$ component can be characterized as paramagnetic materials. It was observed that magnetization values clearly correlate with the chemical composition of powders, and the gradual increase of the BiMnO$_3$ content resulted in noticeably higher magnetization values.

Keywords: SrTiO$_3$; BiMnO$_3$; solid solutions; perovskites; sol-gel method; magnetic properties

1. Introduction

Multiferroics are considered a multifunctional class of compounds, in which at least two ferroic orders (ferroelastic, ferromagnetic, or ferroelectric) exist simultaneously. In this type of material, the manipulation of magnetic properties is possible via the application of an external electric field and controllability of ferroic features can be achieved via the application of an external magnetic field [1,2]. The strong magnetoelectric coupling makes multiferroics useful in different kinds of technological fields, including gyrators [3], magnetic sensors [4], memory devices [5], etc. All multiferroic ceramics can be divided into two main groups: Single-phase compounds and composites. By combining phases responsible for ferroelectric and magnetic properties separately, the electrical, magnetic, and magnetoelectric characteristics can be improved in comparison with single-phase materials [6,7].

One of the most studied single-phase multiferroics is BiMnO$_3$ (BMO) [8]. In the form of a thin film, this material has shown good ferroelectric properties, with remnant polarization of 23 µC/cm^2 at 5 K [9] and 16 µC/cm^2 at room temperature [10]. Moreover, BMO is ferromagnetic at low temperatures in bulk ($T_C = 105$ K) [11] and thin film forms [12]. This perovskite-type material is considered metastable since heating at ambient pressure leads to its decomposition and the formation of Bi$_2$Mn$_4$O$_{10}$ and Bi$_{12}$MnO$_{20}$

phases [13]. Furthermore, for the preparation of BMO, a high pressure is required [13,14], but even with the applied pressure, the formation of monophasic perovskite phase remains a challenging task.

$SrTiO_3$ (STO) is a well-known perovskite-type material, which can be applied in different technological fields, such as thermoelectrics [15], photocatalysis for water splitting [16], piezoelectrics [17], superconductors [18], solid oxide fuel cells [19], etc. This material is considered to be quantum paraelectric, since the ferroelectric ordering in the lower tetragonal symmetry is suppressed by quantum fluctuations [20]. Transition to the ferroelectric phase in this compound can be induced by different methods, including ^{16}O substitution by an ^{18}O isotope [21], cation substitution [22], or the application of intense terahertz electric field excitation [23]. The hydrothermal method [24], polymerized complex method [25], solid-state reaction [26], spray-drying [27], and molten salt [28] are a few of the synthetic approaches that have been previously employed for the synthesis of STO.

To the best of our knowledge, there is only one article regarding powdered (1−x)STO-xBMO solid solutions investigating compositions in the range of 30–70 mol% BMO [29]. It was observed that solid solutions prepared by the solid-state reaction method and having 60 and 70 mol% of BMO exhibited high room-temperature dielectric loss. However, the authors did not show the XRD patterns of their synthesized compounds. Therefore, there is no evidence of the formation of single-phase solid solution in such a broad compositional range. Taking into account the hardly achievable stabilization of materials containing 70 mol% of BMO, those results do not seem reliable. Additionally, perovskite BMO thin films can be synthesized on STO substrates [30]. Furthermore, solid solutions containing both STO and multiferroic perovskite, such as $BiFeO_3$ [31], can be prepared. On the other hand, BMO can be partially stabilized in a solid solution form with $PbTiO_3$ [32] or $LaMnO_3$ [33]. To the best of our knowledge, there are no studies reporting on the application of the sol-gel synthesis of STO-BMO solid solutions.

In the present work, we developed an aqueous sol-gel synthetic approach to prepare (1−x)$SrTiO_3$-x$BiMnO_3$ ((1−x)STO-xBMO) solid solutions for the very first time. The maximal substitution level, structural, morphological, and magnetic properties were investigated for all synthesized materials.

2. Experiment

2.1. Synthesis

For the synthesis of (1−x)STO-xBMO solid solutions, strontium nitrate ($Sr(NO_3)_2$, ≥99.0%, Sigma-Aldrich, Darmstadt, Germany), titanium (IV) isopropoxide ($C_{12}H_{28}O_4Ti$, ≥97%, Sigma-Aldrich, Darmstadt, Germany), bismuth (III) nitrate pentahydrate ($Bi(NO_3)_3·5H_2O$, 98%, Roth, Karlsruhe, Germany), and manganese (II) nitrate tetrahydrate ($Mn(NO_3)_2·4H_2O$, 99.9%, Alfa Aesar, Kandel, Germany) were used as precursors. For the typical synthesis of 1 g of solid solution with a 0.9STO-0.1BMO composition, 6.418 g of citric acid monohydrate ($C_6H_8O_7·H_2O$, 99.9%, Chempur, Piekary Slaskie, Poland) was firstly dissolved in 20 mL of distilled water (the ratio between total metal ions and citric acid was 1:3). Then, 1.360 mL of titanium isopropoxide was added and the hot plate temperature was adjusted to 90 °C. After mixing, a clear and transparent solution was obtained. In the next stage, 0.128 g of $Mn(NO_3)_2·4H_2O$ and 0.970 g of $Sr(NO_3)_2$ were added to the above solution. Prior to the addition of bismuth (III) nitrate (0.2471 g), the pH value of the reaction mixture was adjusted to 1 by adding nitric acid (HNO_3, 65%). This was done in order to avoid the formation of insoluble bismuth oxynitrate $BiONO_3$ precipitates. After that, 5.960 mL of ethylene glycol ($C_2H_6O_2$, Sigma-Aldrich, ≥99.5%) was added to the transparent solution (the ratio between total metal ions and ethylene glycol was 1:10) and the mixture was homogenized at 90 °C for 1.5 h. For the formation of precursor gel, the evaporation of water was performed at 150 °C. The obtained gel was dried for 12 h at 180 °C, and the resulting powder was ground in a mortar and annealed at 1000 °C for 5 h in air atmosphere with a heating rate of 5 °C/min. For the synthesis of solid solutions of other chemical compositions, the required amount of starting materials was recalculated according to the stoichiometry of the final product.

2.2. Characterization

Thermogravimetric and differential scanning calorimetry (TG-DSC) analysis using an STA 6000 Simultaneous Thermal Analyzer (Perkin Elmer, Waltham, MA, USA) was applied to investigate the thermal decomposition of precursor gels. A typical amount of 5–10 mg of dried sample was heated from 30 to 900 °C at a 10 °C/min heating rate in dry flowing air (20 mL/min). A MiniFlex II diffractometer (Rigaku, The Woodlands, TX, USA) working in Bragg–Brentano ($\Theta/2\Theta$) geometry was used for powder X-ray diffraction analysis (XRD). The data were obtained within a 2Θ range from 20 to 80°, with a scanning speed of 5°/min and a step width of 0.02°. Lattice parameters were refined by the Rietveld method using the FullProf suite. Infrared (FT-IR) spectra were obtained in the range of 4000–400 cm^{-1} employing an ALPHA ATR (Bruker, Billerica, Ma, USA) spectrometer. The morphological features of the synthesized products were examined with an SU-70 field-emission scanning electron microscope (FE-SEM, Hitachi, Tokyo, Japan). ImageJ software (LOCI, Madison, WI, USA) was used to estimate the particle size distribution from SEM images. For magnetic measurements, powdered sample (100 mg) was placed in a plastic straw with a 5 mm diameter with a sample height of approximately 5 mm and embedded between foam plugs. The dependence of the magnetization of samples on the strength of the magnetic field was recorded using a magnetometer consisting of an SR510 lock-in amplifier (Stanford Research Systems, Gainesville, GA, USA), an FH-54 Gauss/Teslameter (Magnet Physics, Cologne, Germany), and a laboratory magnet supplied by an SM 330-AR-22 power source (Delta Elektronika, Eindhoven, The Netherlands).

3. Results and Discussion

A thermogravimetric analysis of precursor gels of two different compositions was performed in order to determine the possible temperature of the formation of solid-solutions and to investigate the thermal decomposition behavior. TG-DTG-DSC curves of the gel corresponding to the final STO composition are depicted in Figure 1. It can be seen that thermal degradation of the Sr-Ti-O gel can be divided into three main steps. The first step can be associated with the loss of absorbed water, during which only an insignificant amount of the initial weight (2–3%) is lost. This process occurs in the range of approximately 100–150 °C. The second weight loss takes place at around 300 °C, and this step can be attributed to the decomposition of the organic framework of the gel. Finally, the last weight loss can be seen in the 450–500 °C temperature range. This step is accompanied by a strong exothermic peak, which is clearly seen from the DSC curve. The third stage can be ascribed to the combustion reaction between nitrates and residual organic parts of the gel. The residual mass remains constant at temperatures above 540 °C. The total weight loss calculated from the TG curve was 82%.

For comparison, the thermal decomposition of the gel with the highest BMO content (0.5STO-0.5BMO) was also studied (Figure 2). The degradation behavior is very similar to that of the gel of pristine STO, so it can also be considered a three-step process, and the final weight loss is 80%. The remarkable feature is that the final exothermic degradation step occurs at a slightly lower temperature and the residual mass remains constant at temperatures above approximately 480 °C. This difference can be associated with the enhanced content of nitrate ions, which appear in the gel with appropriate metal precursors, instead of titanium isopropoxide.

Based on the results of the thermal analysis (TG curve), it can be suggested that the minimal annealing temperature for the formation of the mixed metal oxides is 460 °C. However, XRD analysis revealed that the preparation of (1−x)STO xBMO solid solutions at such a low temperature was impossible, since the obtained powders were amorphous and at higher temperatures, polyphasic products were formed. Therefore, a 1000 °C temperature was required for the successful synthesis of highly crystalline materials. Figure 3 represents the XRD patterns of STO-BMO solid solutions.

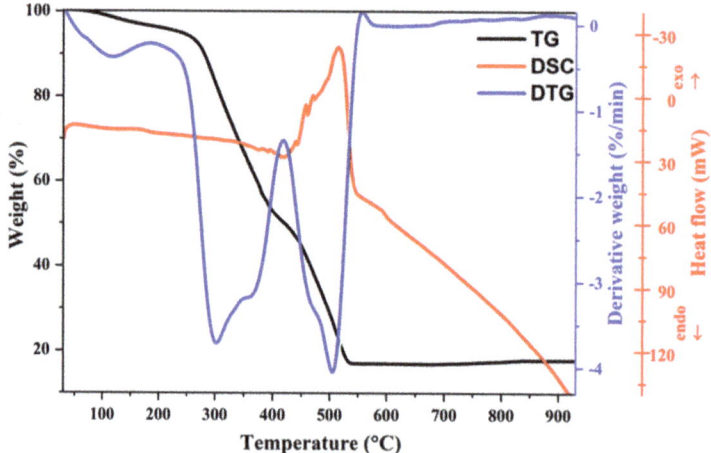

Figure 1. TG-DTG-DSC curves of Sr-Ti-O gel.

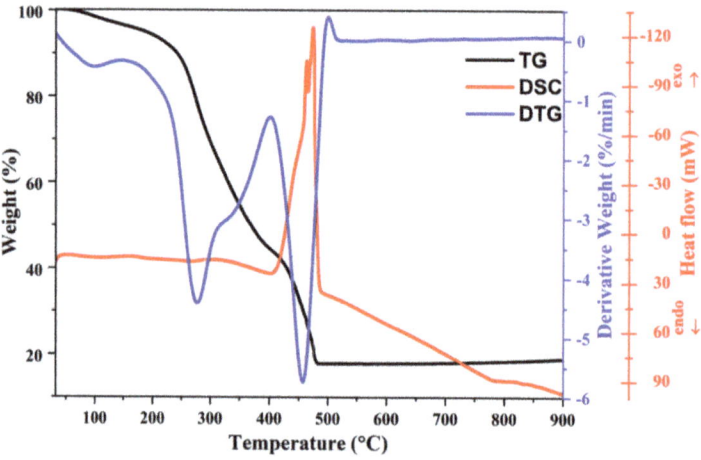

Figure 2. TG-DTG-DSC curves of 0.5(Sr-Ti-O)-0.5(Bi-Mn-O) gel.

It is evident that after the thermal treatment, the pristine STO sample can be characterized as monophasic perovskite material and no traces of impurity phases can be detected. According to the position and intensity of the reflection peaks, the XRD profile matches standard XRD data of $SrTiO_3$ very well (COD #96-900-6865; space group Pm3m; lattice constant a = 3.90528 Å). By employing Rietveld refinement, the cell parameter a = 3.90341 ± 0.00004 Å was calculated and a cubic structure with the space group Pm3m for the pure STO sample was determined. These results are in good agreement with previously reported studies on thin film [34] and bulk [35] STO. The introduction of BMO into the STO structure did not result in any significant structural changes since there is no visible peak splitting or the appearance of extra peaks associated with anything other than perovskite phase. The samples with a higher BMO content (40 and 50 mol%) have a negligible amount of neighboring Mn_3O_4 phase (COD# 96-151-4241). The amount of secondary phase was determined from Rietveld refinement to be 5 and 6% for 0.6STO-0.4BMO and 0.5STO-0.5BMO samples, respectively. Moreover, there was no considerable change in the peak position, despite the different amount of BMO in solid solutions. Similar results were observed in our previous work, where BMO did not cause any visible change in the XRD patterns when introduced into a BaTiO$_3$ crystal structure [36]. Since Bi^{3+} is smaller than Sr^{2+} (1.17 Å versus 1.26 Å in VIII fold coordination) and Mn^{3+} is larger than Ti^{4+} (0.645 Å versus 0.605 Å in VI fold coordination) [37], it is believed that the mismatch in both A and B site elements compensate each other. The products obtained with a BMO content higher than 50 mol% contained a noticeably higher amount of crystalline impurities.

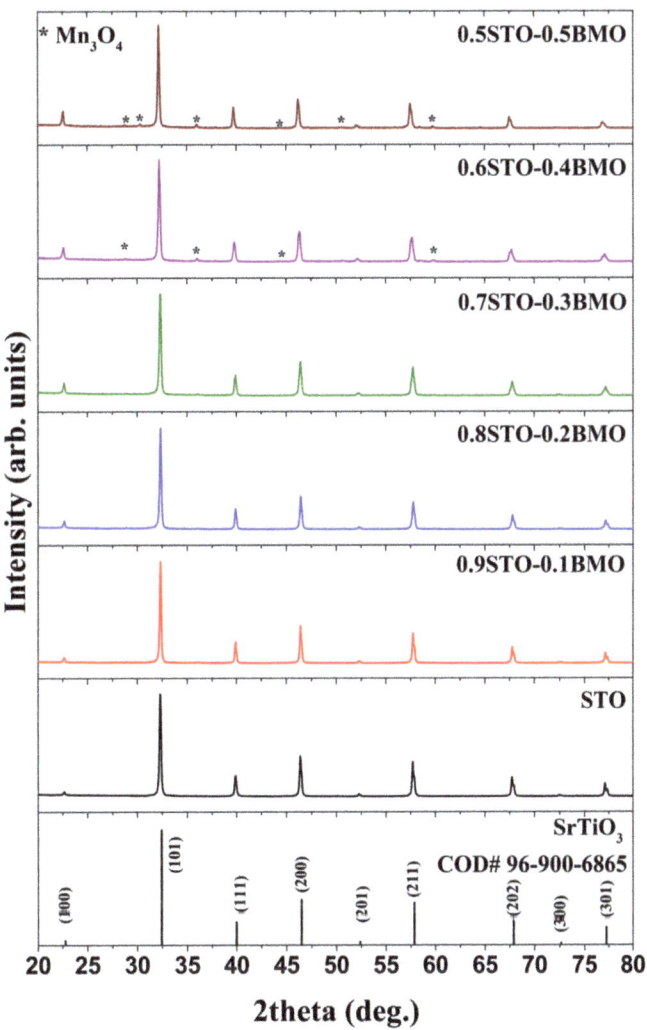

Figure 3. XRD patterns of (1−x)SrTiO$_3$ (STO)-xBiMnO$_3$ (BMO) solid solutions annealed at 1000 °C.

For further structural studies, FT-IR spectroscopy was performed for all samples. The FT-IR spectra of (1−x)STO-xBMO solid solutions are represented in Figure 4. It is known that the FT-IR spectral region of 600–100 cm^{-1} contains three types of vibrations for cubic SrTiO$_3$ perovskite: Ti-O bond stretching vibration at 555 cm^{-1}; Ti-O-Ti bending vibration at 185 cm^{-1}; and Sr-TiO$_3$ lattice mode at 100 cm^{-1} [38]. For our synthesized pristine STO sample, only one broad peak can be observed at 554 cm^{-1}, which can be attributed to Ti-O stretching vibration. With an increasing amount of BMO, the peak center gradually shifts to higher wavenumbers and reaches 622 cm^{-1} for the 0.5STO-0.5BMO sample. According to previous studies, the FT-IR spectrum of phase-pure BMO has two sharp peaks attributed to Mn-O stretching vibrations at 827 and 722 cm^{-1} [39]. The observed peak shift is the result of the superposition effect of Mn-O and Ti-O bond lengths. In pure STO, Ti^{4+} cations and 6 O^{2-} anions form TiO$_6$ octahedra. When introducing Mn^{3+} into the perovskite structure instead of Ti^{4+}, Mn^{3+} cations can form MnO$_6$ octahedra. This substitution causes the changes in the average M-O (where M is Ti^{4+} or Mn^{3+}) distances in the octahedra. Furthermore, there are no absorption peaks belonging to the impurity phase of Mn$_3$O$_4$, which should be located at 631, 529, and 416 cm^{-1}, confirming that the amount of this minor phase is negligible [40].

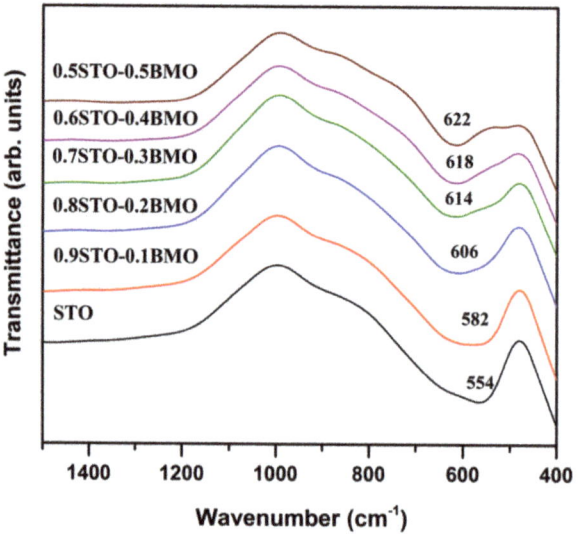

Figure 4. FT-IR spectra of (1−x)STO-xBMO solid solutions annealed at 1000 °C.

The morphology of synthesized (1−x)STO-xBMO solid solutions was examined using SEM. Figure 5 demonstrates the SEM micrographs of STO and representative STO-BMO powders of different compositions. It can be seen that STO powders are composed of mostly uniform and agglomerated nanosized particles, and the size of the particles varies in the range of approximately 40–100 nm. The partial substitution of STO by BMO unequivocally stimulates the significant growth of grains. The sample containing 20 mol% of BMO displays clearly larger particles, and grains with a size of approximately 1–2 µm can be seen. On the other hand, some smaller particles of 0.25–0.6 µm can also be noticed in this sample.

A further increase in the percentage of the BMO component to 40 mol% resulted in further growth of the grains to 1.2–3 µm. Finally, the largest grains ranging from 1.4 to 5 µm were observed for the 0.5STO-0.5BMO sample. It is known that impurity phases in comparison with major phases can possess a completely different morphology and can be often observed in the grain boundary region [41]. In our case, for samples with a minor Mn$_3$O$_4$ phase (Figure 5c,d), we cannot see the formation of particles with obviously different shapes or sizes.

Dependence of the magnetization on the applied magnetic field strength and chemical composition of the solid solutions was further investigated. Room temperature magnetization curves of all samples are represented in Figure 6.

At room temperature, a pure STO compound is considered to be diamagnetic [42]. Ferromagnetism in pristine STO at room temperature was previously only observed in samples with oxygen-deficiency [43] or by induction with laser annealing [44]. From Figure 6, it can be clearly seen that all BMO-containing samples can be characterized by paramagnetic behavior. For paramagnets, magnetization is proportional to the applied magnetic field m = χH. According to Curie–Weiss law [45,46], the molar susceptibility of samples can be described as follows:

$$\chi = \frac{C}{T - T_c} = \frac{n}{3k_B} \cdot \frac{\mu_{eff}^2}{T - T_c} \qquad (1)$$

The inclination of curves (Figure 6a) is defined by BMO molar contribution x, effective magnetic moment of Mn atom μ_{eff}, and Weiss temperature Θ, which depend on the magnetic ordering type and temperature, and thus on the composition and structure of the samples. N_A is the Avogadro number and χ_0 is the small temperature independent term. An increase of the amount of BMO in solid solutions results in a gradual increase in magnetization. On the other hand, the Mn_3O_4 phase is also known to be paramagnetic at room temperature [47,48]. This means that the Mn_3O_4 impurity phase can contribute to the paramagnetism of samples with a higher BMO content (40 and 50 mol%). The data in Figure 6b were compared with the dependences of Equation (1), assuming previously determined values of $\Theta \approx 126$ K and $\mu_{eff} = 4.91$ μ_B for BMO [46]. The effective magnetic moment $\mu_{eff} = g\sqrt{S(S+1)}\mu_B = 4.9\mu_B$, where the gyromagnetic ratio g = 2 and spin S = 2, is characteristic of Mn^{3+}. However, for BMO nanoparticles, lower values of $\mu_{eff} \leq 4.32$ μ_B and $\Theta \leq 70$ K were obtained [49]. The smaller effective magnetic moment can be explained by the potential contribution of Mn^{4+}. The Weiss temperature Θ is expected to decrease with a decrease in the amount of BMO in solid solution, as Mn is more diluted, decreasing the exchange interaction strength.

Due to the contribution of Mn_3O_4, the susceptibility and amount of Mn in STO-BMO solutions were corrected. The molar contribution of Mn_3O_4 y = 0.05 and 0.06 for 0.6STO-0.4BMO and 0.5STO-0.5BMO samples, respectively, adds to the total susceptibility of samples, according to $\chi = (1-y)\chi^{STO-BMO} + y\chi^{Mn_3O_4}$, where $\chi^{Mn_3O_4}$ is the susceptibility of Mn_3O_4 [48] and $\chi^{STO-BMO}$ is that of STO-BMO solid solution. The molar contribution of Mn in STO-BMO solid solutions $\chi^{STO-BMO}$ decreases because of the formation of Mn_3O_4 as the amount of Mn in both phases is $\chi = (1-y)\chi^{STO-BMO} + 3y$. The corrected values (red points in Figure 6b) fit the line obtained with $\mu_{eff} = 4.9\mu_B$ for Mn^{3+} in solid solutions relatively well, however, considering that a composition with smaller x should result in a lower Θ.

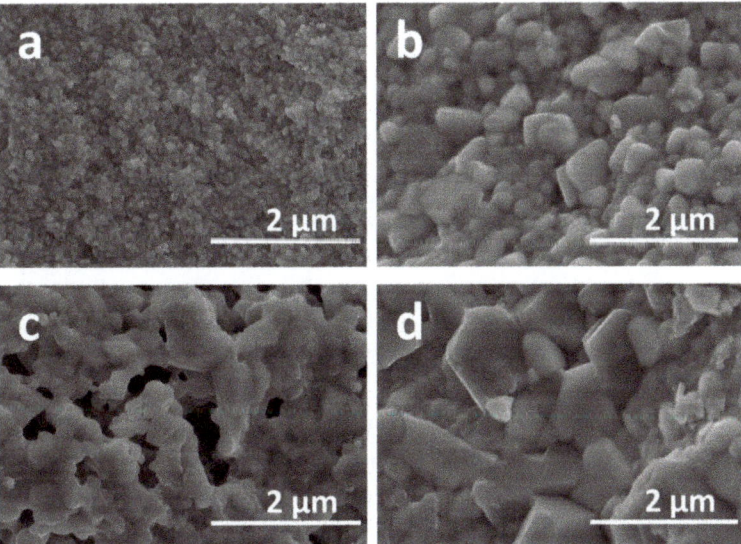

Figure 5. SEM micrographs of STO. (**a**) 0.8STO-0.2BMO, (**b**) 0.6STO-0.4BMO (**c**), and 0.5STO-0.5BMO (**d**) solid solutions annealed at 1000 °C.

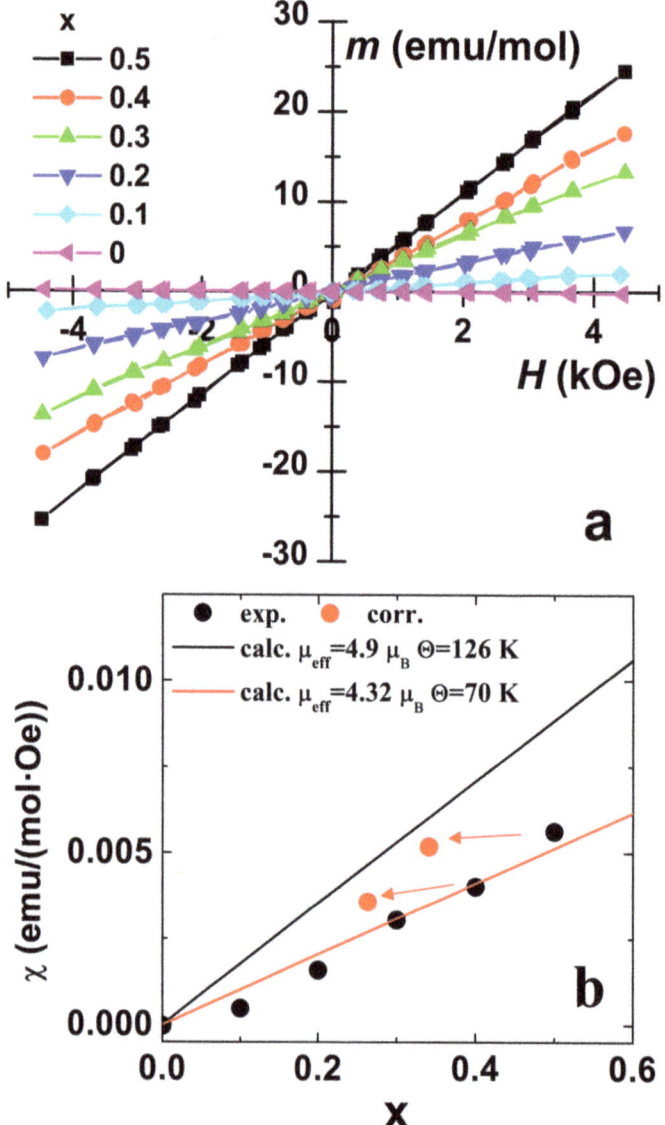

Figure 6. Magnetization curves of (1−x)STO-xBMO solid solutions annealed at 1000 °C (a) and a comparison of calculated and experimental molar susceptibility values (b).

4. Conclusions

In this work, (1−x)SrTiO$_3$-xBiMnO$_3$ ((1−x)STO-xBMO) solid solutions were synthesized for the first time, employing an aqueous sol-gel method. The single-phase perovskites were obtained in the compositional range of 0 to 30 mol% of BMO. The formation of a negligible amount of neighboring Mn$_3$O$_4$ impurity phase was observed for the samples with a higher amount of BMO. FT-IR analysis of synthesized specimens showed gradual spectral change associated with the superposition effect of Mn-O and Ti-O bond lengths with an increase of the BMO content in the solid solutions. The introduction of BMO into STO led to a considerable increase in the grain size. The magnetic properties of the samples were determined to be strongly dependent on the chemical composition of the samples. Paramagnetic behavior was observed for all BMO-containing samples and the magnetization values increased with an increase in the amount of BMO. It was demonstrated that the sol-gel synthetic approach is a suitable method for the successful preparation of $BiMnO_3$-containing mixed metal oxides. This work can be considered as a starting point in the investigation of the structural and physical properties of STO-BMO solid solutions.

Author Contributions: Conceptualization, D.K. and A.K.; methodology, D.K., R.D., and A.Z.; formal analysis, D.K., R.D., E.R.-S., K.M., and D.B.; investigation, D.K., R.D., E.R.-S., K.M., D.B., and A.Z.; resources, A.Z., A.B., and A.K.; writing—original draft preparation, D.K.; writing—review and editing, A.K.; supervision, A.B. and A.K.; funding acquisition, A.Z. and A.K. All authors have read and agreed to the published version of the manuscript.

Funding: This work was supported by the research grant BUNACOMP (No. S-MIP-19-9) from the Research Council of Lithuania. The World Federation of Scientists is highly acknowledged for a National Scholarship awarded to A.Z.

Conflicts of Interest: The authors declare no conflict of interest.

References

1. Zavaliche, F.; Zheng, H.; Mohaddes-Ardabili, L.; Yang, S.Y.; Zhan, Q.; Shafer, P.; Reilly, E.; Chopdekar, R.; Jia, Y.; Wright, P.; et al. Electric Field-Induced Magnetization Switching in Epitaxial Columnar Nanostructures. *Nano Lett.* **2005**, *5*, 1793–1796. [CrossRef] [PubMed]
2. Evans, D.M.; Schilling, A.; Kumar, A.; Sanchez, D.; Ortega, N.; Arredondo, M.; Katiyar, R.S.; Gregg, J.M.; Scott, J.F. Magnetic switching of ferroelectric domains at room temperature in multiferroic PZTFT. *Nat. Commun.* **2013**, *4*, 1534. [CrossRef] [PubMed]
3. Leung, C.M.; Sreenivasulu, G.; Zhuang, X.; Tang, X.; Gao, M.; Xu, J.; Li, J.; Zhang, J.; Srinivasan, G.; Viehland, D. A Highly Efficient Self-Biased Nickel-Zinc Ferrite/Metglas/PZT Magnetoelectric Gyrator. *Phys. Status Solidi (RRL) Rapid Res. Lett.* **2018**, *12*, 1800043. [CrossRef]
4. Marauska, S.; Jahns, R.; Greve, H.; Quandt, E.; Knöchel, R.; Wagner, B. MEMS magnetic field sensor based on magnetoelectric composites. *J. Micromech. Microeng.* **2012**, *22*, 65024. [CrossRef]
5. Bur, A.; Wong, K.; Zhao, P.; Lynch, C.S.; Amiri, P.K.; Wang, K.L.; Carman, G.P. Electrical control of reversible and permanent magnetization reorientation for magnetoelectric memory devices. *Appl. Phys. Lett.* **2011**, *98*, 262504. [CrossRef]
6. Jain, A.; Wang, Y.; Wang, N.; Li, Y.; Wang, F. Emergence of ferrimagnetism along with magnetoelectric coupling in $Ba_{0.83}Sr_{0.07}Ca_{0.10}TiO_3/BaFe_{12}O_{19}$ multiferroic composites. *J. Alloys Compd.* **2020**, *818*, 152838. [CrossRef]
7. Jain, A.; Wang, Y.; Wang, N.; Li, Y.; Wang, F. Existence of heterogeneous phases with significant improvement in electrical and magnetoelectric properties of $BaFe_{12}O_{19}/BiFeO_3$ multiferroic ceramic composites. *Ceram. Int.* **2019**, *45*, 22889–22898. [CrossRef]
8. Belik, A.A. Polar and nonpolar phases of $BiMO_3$: A review. *J. Solid State Chem.* **2012**, *195*, 32–40. [CrossRef]
9. Jeen, H.; Singh-Bhalla, G.; Mickel, P.R.; Voigt, K.; Morien, C.; Tongay, S.; Hebard, A.F.; Biswas, A. Growth and characterization of multiferroic $BiMnO_3$ thin films. *J. Appl. Phys.* **2011**, *109*, 74104. [CrossRef]
10. Son, J.Y.; Shin, Y.H. Multiferroic $BiMnO_3$ thin films with double $SrTiO_3$ buffer layers. *Appl. Phys. Lett.* **2008**, *93*, 62902. [CrossRef]
11. Dos Santos, A.M.; Cheetham, A.K.; Atou, T.; Syono, Y.; Yamaguchi, Y.; Ohoyama, K.; Chiba, H.; Rao, C.N.R. Orbital ordering as the determinant for ferromagnetism in biferroic $BiMnO_3$. *Phys. Rev. B* **2002**, *66*, 064425. [CrossRef]
12. De Luca, G.M.; Preziosi, D.; Chiarella, F.; Di Capua, R.; Gariglio, S.; Lettieri, S.; Salluzzo, M. Ferromagnetism and ferroelectricity in epitaxial $BiMnO_3$ ultra-thin films. *Appl. Phys. Lett.* **2013**, *103*, 062902. [CrossRef]
13. Montanari, E.; Righi, L.; Calestani, G.; Migliori, A.; Gilioli, E.; Bolzoni, F. Room Temperature Polymorphism in Metastable $BiMnO_3$ Prepared by High-Pressure Synthesis. *Chem. Mater.* **2005**, *17*, 1765–1773. [CrossRef]
14. Toulemonde, P.; Darie, C.; Goujon, C.; Legendre, M.; Mendonça, T.; Álvarez-Murga, M.; Simonet, V.; Bordet, P.; Bouvier, P.; Kreisel, J.; et al. Single crystal growth of $BiMnO_3$ under high pressure–high temperature. *High Press. Res.* **2009**, *29*, 600–604. [CrossRef]

15. Muta, H.; Kurosaki, K.; Yamanaka, S. Thermoelectric properties of rare earth doped SrTiO$_3$. *J. Alloys Compd.* **2003**, *350*, 292–295. [CrossRef]
16. Iwashina, K.; Kudo, A. Rh-Doped SrTiO$_3$Photocatalyst Electrode Showing Cathodic Photocurrent for Water Splitting under Visible-Light Irradiation. *J. Am. Chem. Soc.* **2011**, *133*, 13272–13275. [CrossRef]
17. Guo, Y.; Kakimoto, K.-I.; Ohsato, H. Dielectric and piezoelectric properties of lead-free (Na$_{0.5}$K$_{0.5}$)NbO$_3$–SrTiO$_3$ ceramics. *Solid State Commun.* **2004**, *129*, 279–284. [CrossRef]
18. Ahadi, K.; Galletti, L.; Li, Y.; Salmani-Rezaie, S.; Wu, W.; Stemmer, S. Enhancing superconductivity in SrTiO$_3$ films with strain. *Sci. Adv.* **2019**, *5*, eaaw0120. [CrossRef]
19. Hui, S.; Petric, A. Evaluation of yttrium-doped SrTiO$_3$ as an anode for solid oxide fuel cells. *J. Eur. Ceram. Soc.* **2002**, *22*, 1673–1681. [CrossRef]
20. Rowley, S.E.; Spalek, L.J.; Smith, R.P.; Dean, M.P.M.; Itoh, M.; Scott, J.F.; Lonzarich, G.G.; Saxena, S.S. Ferroelectric quantum criticality. *Nat. Phys.* **2014**, *10*, 367–372. [CrossRef]
21. Itoh, M.; Wang, R.; Inaguma, Y.; Yamaguchi, T.; Shan, Y.J.; Nakamura, T. Ferroelectricity Induced by Oxygen Isotope Exchange in Strontium Titanate Perovskite. *Phys. Rev. Lett.* **1999**, *82*, 3540–3543. [CrossRef]
22. Lemanov, V.V.; Smirnova, E.P.; Syrnikov, P.P.; Tarakanov, E.A. Phase transitions and glasslike behavior in Sr$_{1-x}$Ba$_x$TiO$_3$. *Phys. Rev. B* **1996**, *54*, 3151–3157. [CrossRef] [PubMed]
23. Li, X.; Qiu, T.; Zhang, J.; Baldini, E.; Lu, J.; Rappe, A.M.; Nelson, K.A. Terahertz field–induced ferroelectricity in quantum paraelectric SrTiO$_3$. *Science* **2019**, *364*, 1079–1082. [CrossRef]
24. Zhang, S.; Liu, J.; Han, Y.; Chen, B.; Li, X. Formation mechanisms of SrTiO$_3$ nanoparticles under hydrothermal conditions. *Mater. Sci. Eng. B* **2004**, *110*, 11–17. [CrossRef]
25. Chen, W.; Liu, H.; Li, X.; Liu, S.; Gao, L.; Mao, L.; Fan, Z.; Shangguan, W.; Fang, W.; Liu, Y. Polymerizable complex synthesis of SrTiO$_3$:(Cr/Ta) photocatalysts to improve photocatalytic water splitting activity under visible light. *Appl. Catal. B Environ.* **2016**, *192*, 145–151. [CrossRef]
26. Silva, E.R.; Curi, M.; Furtado, J.; Ferraz, H.; Secchi, A.R. The effect of calcination atmosphere on structural properties of Y-doped SrTiO$_3$ perovskite anode for SOFC prepared by solid-state reaction. *Ceram. Int.* **2019**, *45*, 9761–9770. [CrossRef]
27. Duong, H.P.; Mashiyama, T.; Kobayashi, M.; Iwase, A.; Kudo, A.; Asakura, Y.; Yin, S.; Kakihana, M.; Kato, H. Z-scheme water splitting by microspherical Rh-doped SrTiO$_3$ photocatalysts prepared by a spray drying method. *Appl. Catal. B Environ.* **2019**, *252*, 222–229. [CrossRef]
28. Liu, Y.; Qian, Q.; Li, J.; Zhu, X.; Zhang, M.; Zhang, T. Photocatalytic Properties of SrTiO$_3$ Nanocubes Synthesized Through Molten Salt Modified Pechini Route. *J. Nanosci. Nanotechnol.* **2016**, *16*, 12321–12325. [CrossRef]
29. Shevchuk, Y.A.; Shevchuk, Y.A.; Korchagina, S.K.; Ivanova, V.V. Dielectric and Magnetic Properties of SrTiO$_3$–BiMnO$_3$ Solid Solutions. *Inorg. Mater.* **2004**, *40*, 292–294. [CrossRef]
30. Salluzzo, M.; Gariglio, S.; Stornaiuolo, D.; Sessi, V.; Rusponi, S.; Piamonteze, C.; De Luca, G.M.; Minola, M.; Marré, D.; Gadaleta, A.; et al. Origin of Interface Magnetism in BiMnO$_3$/SrTiO$_3$ LaAlO$_3$/SrTiO$_3$ Heterostructures. *Phys. Rev. Lett.* **2013**, *111*, 087204. [CrossRef]
31. Lu, L.; Lv, M.; Liu, G.; Xu, X. Photocatalytic hydrogen production over solid solutions between BiFeO$_3$ and SrTiO$_3$. *Appl. Surf. Sci.* **2017**, *391*, 535–541. [CrossRef]
32. Woodward, D.I.; Reaney, I.M. A structural study of ceramics in the (BiMnO$_3$)$_x$–(PbTiO$_3$)$_{1-x}$ solid solution series. *J. Phys. Condens. Matter* **2004**, *16*, 8823–8834. [CrossRef]
33. Karoblis, D.; Mazeika, K.; Baltrunas, D.; Lukowiak, A.; Strek, W.; Zarkov, A.; Kareiva, A. Novel synthetic approach to the preparation of single-phase Bi$_x$La$_{1-x}$MnO$_{3+\delta}$ solid solutions. *J. Sol-Gel Sci. Technol.* **2019**, *93*, 650–656. [CrossRef]
34. Sharma, D.; Upadhyay, S.; Satsangi, V.R.; Shrivastav, R.; Waghmare, U.V.; Dass, S. Improved Photoelectrochemical Water Splitting Performance of Cu$_2$O/SrTiO$_3$ Heterojunction Photoelectrode. *J. Phys. Chem. C* **2014**, *118*, 25320–25329. [CrossRef]
35. Kim, S.; Choi, H.; Lee, M.; Park, J.; Kim, D.; Do, D.; Kim, M.; Song, T.K.; Kim, W. Electrical properties and phase of BaTiO$_3$–SrTiO$_3$ solid solution. *Ceram. Int.* **2013**, *39*, S487–S490. [CrossRef]
36. Karoblis, D.; Zarkov, A.; Mazeika, K.; Baltrunas, D.; Niaura, G.; Beganskiene, A.; Kareiva, A. Sol-gel synthesis, structural, morphological and magnetic properties of BaTiO$_3$–BiMnO$_3$ solid solutions. *Ceram. Int.* **2020**, *46*, 16459–16464. [CrossRef]

37. Shannon, R.D. Revised effective ionic radii and systematic studies of interatomic distances in halides and chalcogenides. *Acta Crystallogr. Sect. A Cryst. Phys. Diffr. Theor. Gen. Crystallogr.* **1976**, *32*, 751–767. [CrossRef]
38. Perry, C.H.; Khanna, B.N.; Rupprecht, G. Infrared Studies of Perovskite Titanates. *Phys. Rev.* **1964**, *135*, A408–A412. [CrossRef]
39. Bhardwaj, N.; Gaur, A.; Yadav, K. Effect of doping on optical properties in $BiMn_{1-x}(TE)_xO_3$ (where x = 0.0, 0.1 and TE = Cr, Fe, Co, Zn) nanoparticles synthesized by microwave and sol-gel methods. *Appl. Phys. A* **2017**, *123*, 429. [CrossRef]
40. Ocaña, M. Uniform particles of manganese compounds obtained by forced hydrolysis of manganese (II) acetate. *Colloid Polym. Sci.* **2000**, *278*, 443–449. [CrossRef]
41. Sinusaite, L.; Popov, A.; Antuzevics, A.; Mazeika, K.; Baltrunas, D.; Yang, J.-C.; Horng, J.L.; Shi, S.; Sekino, T.; Ishikawa, K.; et al. Fe and Zn co-substituted beta-tricalcium phosphate (β-TCP): Synthesis, structural, magnetic, mechanical and biological properties. *Mater. Sci. Eng. C* **2020**, *112*, 110918. [CrossRef] [PubMed]
42. Coey, J.M.D.; Venkatesan, M.; Stamenov, P. Surface magnetism of strontium titanate. *J. Phys. Condens. Matter* **2016**, *28*, 485001. [CrossRef] [PubMed]
43. Trabelsi, H.; Bejar, M.; Dhahri, E.; Sajieddine, M.; Valente, M.A.; Zaoui, A.; Moez, B. Effect of the oxygen deficiencies creation on the suppression of the diamagnetic behaviour of $SrTiO_3$ compound. *J. Alloys Compd.* **2016**, *680*, 560–564. [CrossRef]
44. Rao, S.S.; Lee, Y.F.; Prater, J.T.; Smirnov, A.I.; Narayan, J. Laser annealing induced ferromagnetism in $SrTiO_3$ single crystal. *Appl. Phys. Lett.* **2014**, *105*, 042403. [CrossRef]
45. Kittel, C.; McEuen, P. *Introduction to Solid State Physics*; Wiley: Hoboken, NJ, USA, 1996; Volume 8.
46. Belik, A.A.; Takayama-Muromachi, E. Magnetic Properties of $BiMnO_3$ Studied with Dc and Ac Magnetization and Specific Heat. *Inorg. Chem.* **2006**, *45*, 10224–10229. [CrossRef]
47. Dwight, K.; Menyuk, N. Magnetic Properties of Mn_3O_4 and the Canted Spin Problem. *Phys. Rev.* **1960**, *119*, 1470–1479. [CrossRef]
48. Srinivasan, G.; Seehra, M.S. Magnetic properties of Mn_3O_4 and a solution of the canted-spin problem. *Phys. Rev. B* **1983**, *28*, 1–7. [CrossRef]
49. Stanojević, Z.M.; Branković, Z.; Jagličić, Z.; Jagodič, M.; Mančić, L.; Bernik, S.; Rečnik, A. Structural and magnetic properties of nanocrystalline bismuth manganite obtained by mechanochemical synthesis. *J. Nanopart. Res.* **2011**, *13*, 3431–3439. [CrossRef]

Publisher's Note: MDPI stays neutral with regard to jurisdictional claims in published maps and institutional affiliations.

 © 2020 by the authors. Licensee MDPI, Basel, Switzerland. This article is an open access article distributed under the terms and conditions of the Creative Commons Attribution (CC BY) license (http://creativecommons.org/licenses/by/4.0/).

Communication

Broadband Detection Based on 2D Bi$_2$Se$_3$/ZnO Nanowire Heterojunction

Zhi Zeng [1,2], Dongbo Wang [1,2,*], Jinzhong Wang [1,2,*], Shujie Jiao [1,2,*], Donghao Liu [1,2], Bingke Zhang [1,2], Chenchen Zhao [1,2], Yangyang Liu [1,2], Yaxin Liu [1,2], Zhikun Xu [3,4,*], Xuan Fang [5,*] and Liancheng Zhao [1,2]

1. National Key Laboratory for precision Hot Processing of Metals, Harbin Institute of Technology, Harbin 150001, China; zengzhi@hit.edu.cn (Z.Z.); 19s009009@stu.hit.edu.cn (D.L.); 18S009003@stu.hit.edu.cn (B.Z.); 20b909059@stu.hit.edu.cn (C.Z.); 20s009008@stu.hit.edu.cn (Y.L.); 20s009004@stu.hit.edu.cn (Y.L.); lczhao@hit.edu.cn (L.Z.)
2. Department of Optoelectronic Information Science, Harbin Institute of Technology, School of Materials Science and Engineering, Harbin 150001, China
3. College of Science, Guangdong University of Petrochemical Technology, Maoming City 525000, China
4. Key Laboratory for Photonic and Electronic Bandgap Materials, Ministry of Education, School of Physics and Electronic Engineering, Harbin Normal University, Harbin 150025, China
5. State Key Laboratory of High Power Semiconductor Lasers, School of Science, Changchun University of Science and Technology, Changchun 130022, China
* Correspondence: wangdongbo@hit.edu.cn (D.W.); jinzhong_wang@hit.edu.cn (J.W.); shujiejiao@hit.edu.cn (S.J.); xuzhikun@gdupt.edu.cn (Z.X.); fangx@cust.edu.cn (X.F.)

Abstract: The investigation of photodetectors with broadband response and high responsivity is essential. Zinc Oxide (ZnO) nanowire has the potential of application in photodetectors, owing to the great optoelectrical property and good stability in the atmosphere. However, due to a large number of nonradiative centers at interface and the capture of surface state electrons, the photocurrent of ZnO based photodetectors is still low. In this work, 2D Bi$_2$Se$_3$/ZnO NWAs heterojunction with type-I band alignment is established. This heterojunction device shows not only an enhanced photoresponsivity of 0.15 A/W at 377 nm three times of the bare ZnO nanowire (0.046 A/W), but also a broadband photoresponse from UV to near infrared region has been achieved. These results indicate that the Bi$_2$Se$_3$/ZnO NWAs type-I heterojunction is an ideal photodetector in broadband detection.

Keywords: 2D Bi$_2$Se$_3$/ZnO NWAs heterojunction; broadband detection; ZnO NWAs; UV detection

1. Introduction

Over the past decades, photodetectors have been extensively used in both military and civil fields, such as living cell inspection [1], night vision [2], optical communications [3], atmospheric [4], etc. [5–8]. Photodetectors have become indispensable in daily life. Among these semiconductor materials for making photodetectors, ZnO has the potential of application in UV photodetectors. Owing to the great optoelectrical property and good stability in the atmosphere, ZnO is widely researched with the band gap of 3.37 eV, of which brought by high exciton binding energy (60 meV) at room temperature, noise interference is suppressed. In addition, the merits of being low-cost, non-toxic and easy to prepare make it more attractive to be investigated [9–11]. Various morphologies of ZnO have been prepared during years of research, such as nanoparticle, nanowire, nanotube, nanofilm, and nanosheet [12–15]. The morphology with different dimensions has different characteristics. With the morphology of one-dimensional nanowire arrays (NWAs), not only the space between the nanowire is well backed up for the strengthening of light trapping ability but also the superior transport properties provide the fast electron transport channel when making comparisons with its bulk and thin-film structures [16,17].

However, until now, the photoresponse performance of the ZnO nanowires UV detector is still lower than the theoretical predicted value, due to a large number of nonradiative centers at interface and electron trapped by the surface states [18].

Many methods have been used to optimize the performance of the ZnO nanowires UV photodetector (UVPD). Yang et al. reported photocurrent enhanced through optimizing ZnO seed layer growth condition [19]. Research of Kim et al. shows that NiO/Ni coated ZnO NWs reveal raised D0X transition due to the increasing oxygen deficiency which is responsible for increasing donor density [20]. Zhang et al. reported a two-dimensional graphene/ZnO nanowire mixed-dimensional van der Waals heterostructure for high-performance photosensing [21]. Zang et al. reported enhancing photoresponse based on ZnO nanoparticles decorated $CsPbBr_3$ films [22]. Sumesh et al. reported broadband and highly sensitive photodetector based on ZnO/WS_2 heterojunction [23]. So far, more and more two-dimensional materials/ZnO nanowires mixed dimensional heterojunctions have been reported [24–27].

Up to now, as the discovery of the unique characteristic of topological insulators (TI), further attention is paid to TI to fabricate photodetectors [28–31]. Taking advantage of Dirac dispersion and spin-momentum locking property brought by 2D surface electrons and time-reversal symmetry, back scattering in the Dirac fermions caused by nonmagnetic impurities is prevented, which enable the outstanding transport characteristics, thus reducing the dark current to obtain higher performance [32]. With direct band gap of 0.3 eV and weak Van der Waals' force between each two layers, Bi_2Se_3 has very infusive photoelectric properties, such as tunable surface bandgap, polarization-sensitive photocurrent, and thickness dependent optical absorption [33,34]. These special properties make Bi_2Se_3 promise for binding with ZnO to build high performance photodetector in the UV and visible region. In contrast, there are scarcely any reports about 2D Bi_2Se_3/ZnO nanowire mixed-dimensional detectors.

In this work, hydrothermal synthesized ZnO NWAs is composite with 2D Bi_2Se_3. The broadband photodetector based on Bi_2Se_3/ZnO NWAs heterojunction photodetector is fabricated. Enhanced UV to visible responsivity is realized in the Bi_2Se_3/ZnO NWAs heterojunction photodetector. The mechanism of enhanced response in Bi_2Se_3/ZnO NWAs heterojunction is dealt with in detail through thorough inspections of photoelectric response characteristic combined with Raman scattering measurements, optical properties, and band gap structure.

2. Materials and Methods

2.1. Syntheses of ZnO NWAs

ZnO NWAs were synthesized via a simple hydrothermal method with buffer layer sputtered on fluorine-doped tin oxide (FTO) glass. A thin buffer layer was deposited with RF magnetron sputtering, during which process, the O_2:Ar flow ratio was controlled at 18:42 in 1 Pa under room temperature for 5 min, after which, it was annealed at 400 °C for 1 h. On the basis of buffer layer, an ordinary hydrothermal method was applied for the synthesis of ZnO nanowire. $Zn(AC)_2 \cdot 6H_2O$ and $C_6H_{12}N_4$ (HMT) with an equal amount of 0.0009 mol were added to 30 mL deionized water and stirred for 5 min, respectively. Subsequently, the two aqueous solutions were mixed and stirred for 5 min. Then, the obtained solution was sealed in autocave at 90 °C for 4 h. Afterwards, synthesized sample was washed with deionized water several times and dried in the air. In this way, ZnO NWAs was prepared.

2.2. Syntheses of 2D Bi_2Se_3

2D Bi_2Se_3 is provided in SixCarbon Technology Shenzhen.

2D Bi_2Se_3 was synthesized via a standard chemical vaporous deposition method. The reactions were conducted in a tube furnace with dual heating zone. 0.1 g 99.995% Bi_2O_3 (Bi_2O_3, 6CARBON, Shenzhen, Guangzhou, China) powder was placed in the high temperature zone to heat up to 700 degrees. 0.5 g 99.999% purity Se (Se, 6CARBON,

Shenzhen, Guangzhou, China) particles were placed in the low temperature zone to heat up to 300 degrees. Under mixed carrier gas of argon and hydrogen at flow rates of 200 sccm and 15 sccm, respectively, the clean sapphire film placed 5 cm below Bi_2O_3 was heated to 500 degrees for 15 min.

2.3. The Transfer of 2D Bi_2Se_3

Methyl methacrylate (PMMA) was coated on obtained 2D Bi_2Se_3. After curing, PMMA was placed in pure water and heated to 90 degrees for 1 h. Then, it was quickly plunged into ice water to separate Bi_2Se_3 which attached to PMMA from sapphire base. Then, the film was placed on ZnO NWAs and finally PMMA was removed with acetone to obtain the transferred Bi_2Se_3 film.

2.4. Characterization

The morphologies of ZnO NWAs and Bi_2Se_3 film were conducted on a field-emission scanning electron microscopy (FE-SEM, ZEISS Merlin Compact, Oberkochen, Germany). Surface morphologies of Bi_2Se_3 film were also characterized by atomic force microscopy (AFM, Dimension Fastscan, Bruker, Billerica, MA, USA). Raman were carried out on (LabRAM HR Evolution, Horiba, Paris, France) with an excitation wavelength of 532 nm. Photoluminescence (PL) spectrums were recorded in CCD using the same instrument system with Raman by He–Cd laser line of 325 nm with fixed excitation intensity at room temperature. Both measurement of PL and Raman use the same equipment. The responsivity of samples was investigated by Zolix responsivity measurement system (DSR600, Zolix, Beijing, China), which calibrated via standard silicon cells. The spectral responsivity was measured in terms of the current signal within the range of 300–1000 nm at 4 V bias under room temperature.

3. Results

Figure 1a shows the dramatic structure of the photodetector. Bi_2Se_3 is transferred onto the ZnO nanowires and then onto a steam plate with platinum electrodes. The morphologies of ZnO NWAs and 2D Bi_2Se_3 were further confirmed through SEM and AFM exhibited in Figures S1 and S2 (Supplementary Materials). Due to stress and other reasons, the Bi_2Se_3 split into several pieces after the transfer. The photodetector is made by the standard semiconductor fabrication techniques. The interdigital metal electrodes, which are defined on a 300 nm Pt layer by the conventional UV photolithography and lift-off procedure, are 0.5 mm long and 300 μm wide, with a 200 μm gap. There were 10 fingers in our interdigital structure, 5 up and 5 down [35], and Figure 1b is a physical image of the detector.

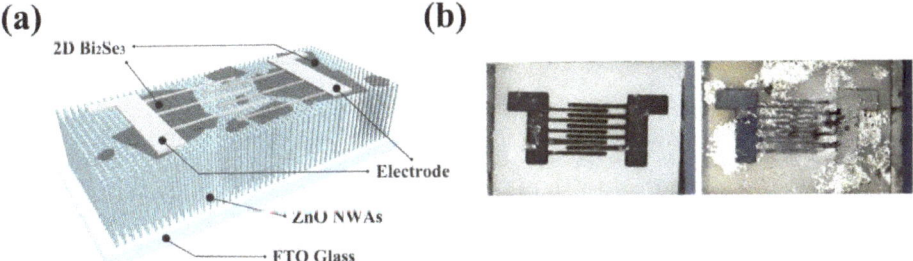

Figure 1. (a) the structure diagram and (b) physical image of photodetector.

As a widely used non-damaged measurement, Raman spectroscopy could be used to investigate intralayer vibration modes, interlayer vibration modes, and the layer coupling in 2D materials effectively [36,37]. Figure 2 showed Raman curves of ZnO NWs, 2D Bi_2Se_3 and Bi_2Se_3/ZnO NWAs. Ascribed to the perpendicular laser incident direction relative to the c-axis of sample surface, there were only two peaks located in 100 cm^{-1} and 438 cm^{-1} corresponding to E_2^{low} mode and E_2^{high} mode of ZnO NWAs in curve (a), respectively, in which the strong E_2^{high} phonon mode peak was observed, indicating the good crystalline quality [38–40]. Located at 72.8 cm^{-1}, 133.3 cm^{-1} and 174.6 cm^{-1}, three peaks observed in curve (b) could be assigned to A_{1g}^1, E_g^2 and A_{1g}^2 vibrational mode in turn for Bi$_2$Se$_3$ [41]. A1g modes were out of plane vibrations and Eg modes were vibrations in-plane [42]. There were two peaks of ZnO NWAs and one peak of Bi$_2$Se$_3$ making their appearance in curve (c), exhibiting the consistent material nature after compositing.

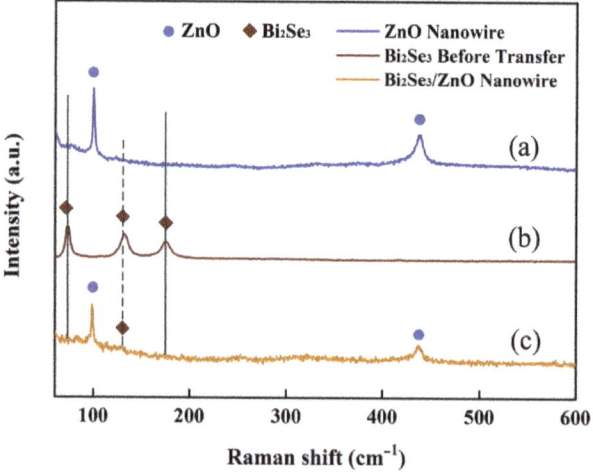

Figure 2. Raman curves of (**a**) ZnO NWAs, (**b**) 2D Bi$_2$Se$_3$ and (**c**) Bi$_2$Se$_3$/ZnO NWAs.

Furthermore, it can be seen in Figure 2, after being transferred onto the ZnO nanowires, the Bi$_2$Se$_3$ mode slightly shifts from 133.3 cm^{-1} to the lower frequency 132 cm^{-1} corresponding to the 2D Bi$_2$Se$_3$. This can be ascribed to the effect of residual stress [43]. The magnitude of the stress can be given by the following formula $\varepsilon = \Delta\omega/\chi$, where χ is the shift rates of Raman vibrational modes. Based on the previous literature reports, the shift rates of E_g^2 mode under biaxial strain is ~5.2 cm^{-1} per % strain [43,44]. Therefore, the stress in the heterojunction should be 0.25%.

Since the photoluminescence characteristic could give an index to the degree of electron and hole recombination of different materials, PL test was carried out on ZnO NWAs and Bi$_2$Se$_3$/ZnO NWAs to find out the change after the heterojunction forming in Figure 3. The sharp emission peak of ZnO NWAs with high intensity at 377 nm derived from near band-edge emissions [45]. In contrast, caused by oxygen vacancies [46] and chemisorbed O$_2$ in the air [47,48], the peak emerging at the visible region is relatively weak, suggesting the high crystal quality. The PL spectrum of Bi$_2$Se$_3$/ZnO NWAs presents an obvious decrease in near band-edge emissions peak in ZnO NWAs, indicating the weakened electron hole recombination. The anomalous decreased PL emission mainly originated from the more efficient separation of photogenerated carriers at the strain-tailored heterointerfaces [43]. In the test of photoluminescence spectrum, we found that ZnO interband luminescence peaks also moved significantly, from 377 nm of simple ZnO nanowires to 378 nm of heterojunctions, which was caused by the band changes brought by stress, according to the relevant literature reports [43]. In the visible light part, the luminous emission peak shifts from 550 nm to 510 nm. It is believed that the radiant peak at 510 nm could be ascribed to the emission of Bi$_2$Se$_3$ [49]. Furthermore, it can be found that after composition with Bi$_2$Se$_3$, the luminescence in the visible region is significantly enhanced, combining with the analysis of the Bi$_2$Se$_3$/ZnO NWAs heterojunction energy band structure below. It can be deduced that Bi$_2$Se$_3$ and ZnO form a type-I band structure, leading to the transfer of photogenerated electrons from ZnO to Bi$_2$Se$_3$. This electron transport process results in the reduction of ZnO emission and the enhancement of Bi$_2$Se$_3$ emission. This will be discussed in detail in the energy band section.

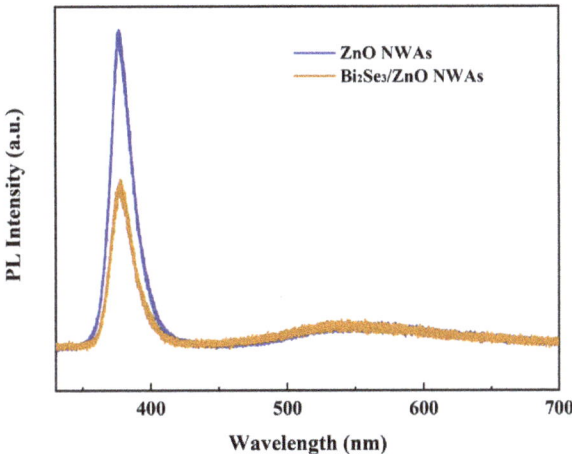

Figure 3. Photoluminescence (PL) spectra of ZnO NWAs and Bi$_2$Se$_3$/ZnO NWAs.

To assess the photoresponse performance of Bi$_2$Se$_3$/ZnO NWAs heterojunction, Pt interdigital electrode was sputtered with a mask to fabricate photodetector, as depicted in Figure 1. Responsivity performance is measured as shown in Figure 4. The spectral responsivity was measured in terms of the current signal in the range of 300–1000 nm at 4 V bias. All the samples revealed UV photoresponse with a cutoff wavelength of 365 nm. This can be attributed to the response of ZnO nanowires. After being transferred with the Bi$_2$Se$_3$, the response of ZnO nanowires enhanced from 0.046 AW^{-1} to 0.15 AW^{-1}, three times higher than the bare ZnO nanowires. Furthermore, the photocurrent of Bi$_2$Se$_3$/ZnO NWAs at visible and near-infrared regions is also increased compared with bare ZnO NWAs. The spectral responses in the visible and near infrared regions are from Bi$_2$Se$_3$, corresponding to the band gap and PL spectrum of Bi$_2$Se$_3$ [50]. The performance of ordinary photodetectors constructed by nanowire and transferred 2D material could be restricted to a large extent. It is difficult to obtain effective responsivity promotion for them as the result of high interface impedance brought by impurity introduced during transfer and the limited contact area between materials of one and two dimensions. Therefore, the performance could be restricted to a large extent. In contrast, the threefold increase responsiveness of the Bi$_2$Se$_3$/ZnO NWAs photodetector shows it is superior to similar detectors owing to the repressed back scattering caused by the intrinsic characteristic of TI [32,51]. In addition, the spectral test shows that the heterojunction not only significantly improves the photoelectric response in the ultraviolet region but also has good spectral response in the visible to near-infrared region, indicating that the heterojunction device has the ability to prepare wide-spectrum detectors.

For ZnO based optoelectronic devices, oxygen molecules adsorbed on the surface of the ZnO nanowires capture the free electrons, due to the surface adsorption and desorption. Therefore, a depletion layer with low conductivity is created near the surface of the film, and results in a reduced photocurrent [52,53].

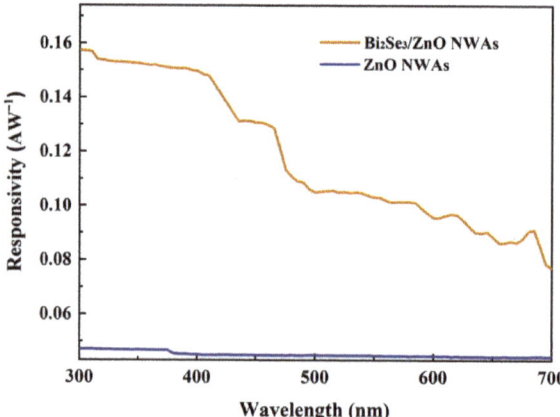

Figure 4. Responsivity spectrum of ZnO NWAs and Bi$_2$Se$_3$/ZnO NWAs photodetector.

In order to study the enhance response mechanism of Bi$_2$Se$_3$/ZnO NWAs heterojunctions, the energy band structure of heterojunctions is analyzed. For the heterojunction, band alignment also plays a crucial role in the spectral response characteristics [54].

The band structure of Bi$_2$Se$_3$/ZnO can be deduced from the Anderson model shown in Figure 5 [55,56]. Based on previously reported energy band data of Bi$_2$Se$_3$ and ZnO [33,34,43,49,50,52,57,58], the conduction band offset (CBO) of the Bi$_2$Se$_3$/ZnO heterojunction is worked out to be 1.67 ± 0.15 eV. With this band alignment, electrons can easily drift from ZnO to the Bi$_2$Se$_3$. As a result, the recombination of electrons and holes may mainly take place on the Bi$_2$Se$_3$ side rather than in ZnO nanowires. This will mitigate the impact of the ZnO surface adsorption and desorption. Hence, at the same time of leading to an increase in the UV current of the Bi$_2$Se$_3$/ZnO NWAs heterojunction, the drifting of electrons also results in a decrease of ZnO emission in the UV region. In this way, the realization of Bi$_2$Se$_3$/ZnO NWAs heterojunction photodetectors makes responsivity of Bi$_2$Se$_3$/ZnO NWAs improve noticeably in the UV region, however, the absorption range broadened effectively.

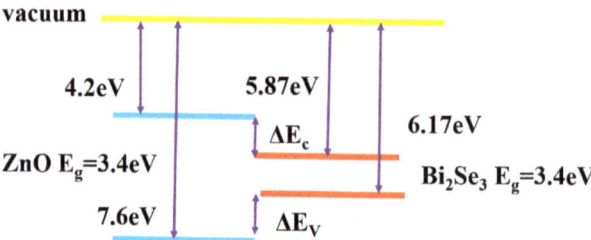

Figure 5. Electronic band alignment for Bi$_2$Se$_3$/ZnO NWAs structure.

4. Conclusions

In summary, this work presents a novel broadband heterojunction photodetector based on 2D Bi$_2$Se$_3$/ZnO NWAs. Combining the merit of 2D Bi$_2$Se$_3$ and ZnO, the responsivity spectrum of the photodetector exhibited the improved responsivity covering UV-Visible-Near Infrared range. The properties of a responsivity of 0.15 A/W, nearly three times to the bare ZnO, benefit from unique advantages of type-I band alignments. More importantly, the special heterojunction photodetector with type-I energy band structure injects a large number of electrons into the ultrathin 2D material to enhance responsivity of Visible-Near

Infrared wavelength while exporting electrons to the electrode rapidly to avoid electron and hole recombination, making it a promising and yet-to-be-extended way for novel exploration for more applications.

Supplementary Materials: The following are available online at https://www.mdpi.com/2073-4352/11/2/169/s1, Figure S1: (a) low magnification and (b) high magnification SEM images of ZnO NWAs; (c) low magnification and (d) high magnification SEM images of Bi_2Se_3; Figure S2: (a) 2D and (b) 3D AFM images of Bi_2Se_3.

Author Contributions: Conceptualization, J.W.; methodology, B.Z., Y.L. (Yaxin Liu), and C.Z.; formal analysis, D.W. and Z.Z.; investigation, D.L.; data curation, S.J. and Y.L. (Yangyang Liu); writing—original draft preparation, Z.Z.; writing—review and editing, B.Z.; supervision, L.Z. and X.F.; funding acquisition, D.W. and Z.X. All authors have read and agreed to the published version of the manuscript.

Funding: This research was funded by National Key Research and Development Program of China, grant number 2019YFA0705201.

Informed Consent Statement: Not applicable.

Data Availability Statement: The data presented in this study are available on request from the corresponding author.

Conflicts of Interest: The authors declare no conflict of interest.

References

1. Bartels, R.A.; Paul, A.; Green, H.; Kapteyn, H.C.; Murnane, M.M.; Backus, S.; Christov, I.P.; Liu, Y.; Attwood, D.; Jacobsen, C. Generation of spatially coherent light at extreme ultraviolet wavelengths. *Science* **2002**, *297*, 376–378. [CrossRef]
2. Huo, N.J.; Konstantatos, G. Recent progress and future prospects of 2d-based photodetectors. *Adv. Mater.* **2018**, *30*, 1801164. [CrossRef]
3. Pospischil, A.; Humer, M.; Furchi, M.M.; Bachmann, D.; Guider, R.; Fromherz, T.; Mueller, T. CMOS-compatible graphene photodetector covering all optical communication bands. *Nat. Photonic.* **2013**, *7*, 892–896. [CrossRef]
4. Formisano, V.; Atreya, S.; Encrenaz, T.; Ignatiev, N.; Giuranna, M. Detection of methane in the atmosphere of Mars. *Science* **2004**, *306*, 1758–1761. [CrossRef]
5. Liu, W.B.; Zhang, J.F.; Zhu, Z.H.; Yuan, X.D.; Qin, S.Q. Electrically tunable absorption enhancement with spectral and polarization selectivity through graphene plasmonic light trapping. *Nanomaterials* **2016**, *6*, 155. [CrossRef] [PubMed]
6. Li, Y.F.; Zhang, Y.T.; Li, T.T.; Li, M.Y.; Chen, Z.L.; Li, Q.Y.; Zhao, H.L.; Sheng, Q.; Shi, W.; Yao, J. Ultrabroadband, ultraviolet to terahertz, and high sensitivity $CH_3NH_3PbI_3$ perovskite photodetectors. *Nano Lett.* **2020**, *20*, 5646–5654. [CrossRef]
7. Aldalbahi, A.; Velazquez, R.; Zhou, A.F.; Rahaman, M.; Feng, P.X. Bandgap-tuned 2d boron nitride/tungsten nitride nanocomposites for development of high-performance deep ultraviolet selective photodetectors. *Nanomaterials* **2020**, *10*, 1433. [CrossRef] [PubMed]
8. Ni, S.M.; Guo, F.Y.; Wang, D.B.; Liu, G.; Xu, Z.K.; Kong, L.P.; Wang, J.Z.; Jiao, S.J.; Zhang, Y.; Yu, Q.J.; et al. Effect of MgO surface modification on the TiO_2 nanowires electrode for self-powered UV photodetectors. *ACS Sustain. Chem. Eng.* **2018**, *6*, 7265–7272. [CrossRef]
9. Liang, S.; Sheng, H.; Liu, Y.; Huo, Z.; Lu, Y.; Shen, H. ZnO Schottky ultraviolet photodetectors. *J. Cryst. Growth* **2001**, *225*, 110–113. [CrossRef]
10. Jiao, S.J.; Zhang, Z.Z.; Lu, Y.M.; Shen, D.Z.; Yao, B.; Zhang, J.Y.; Li, B.H.; Zhao, D.X.; Fan, X.W.; Tang, Z.K. ZnO p-n junction light-emitting diodes fabricated on sapphire substrates. *Appl. Phys. Lett.* **2006**, *88*, 031911. [CrossRef]
11. Monroy, E.; Omnes, F.; Calle, F. Wide-bandgap semiconductor ultraviolet photodetectors. *Semicond. Sci. Technol.* **2003**, *18*, R33–R51. [CrossRef]
12. Jin, Y.Z.; Wang, J.P.; Sun, B.Q.; Blakesley, J.C.; Greenham, N.C. Solution-processed ultraviolet photodetedtors based on colloidal ZnO nanoparticles. *Nano Lett.* **2008**, *8*, 1649–1653. [CrossRef]
13. Gedamu, D.; Paulowicz, I.; Kaps, S.; Lupan, O.; Wille, S.; Haidarschin, G.; Mishra, Y.K.; Adelung, R. Rapid fabrication technique for interpenetrated ZnO nanotetrapod networks for Fast UV sensors. *Adv. Mater.* **2014**, *26*, 1541–1550. [CrossRef] [PubMed]
14. Hu, L.F.; Yan, J.; Liao, M.Y.; Xiang, H.J.; Gong, X.G.; Zhang, L.D.; Fang, X.S. An optimized ultraviolet-a light photodetector with wide-range photoresponse based on ZnS/ZnO biaxial nanobelt. *Adv. Mater.* **2012**, *24*, 2305–2309. [CrossRef]
15. Zhang, B.K.; Li, Q.; Wang, D.B.; Wang, J.Z.; Jiang, B.J.; Jiao, S.J.; Liu, D.H.; Zeng, Z.; Zhao, C.C.; Liu, Y.X.; et al. Efficient photocatalytic hydrogen evolution over TiO_{2-x} mesoporous spheres-ZnO nanowires heterojunction. *Nanomaterials* **2020**, *10*, 2096. [CrossRef]
16. Liu, X.; Gu, L.L.; Zhang, Q.P.; Wu, J.Y.; Long, Y.Z.; Fan, Z.Y. All-printable band-edge modulated ZnO nanowire photodetectors with ultra-high detectivity. *Nat. Commun.* **2014**, *5*, 1–9. [CrossRef] [PubMed]

17. Kim, H.; Yan, H.Q.; Messer, B.; Law, M.; Yang, P.D. Nanowire ultraviolet photodetectors and optical switches. *Adv. Mater.* **2002**, *14*, 158–160.
18. Yang, J.L.; Liu, K.W.; Shen, D.Z. Recent progress of ZnMgO ultraviolet photodetector. *Chin. Phys. B* **2017**, *26*, 047308. [CrossRef]
19. Gu, P.; Zhu, X.H.; Yang, D.Y. Vertically aligned ZnO nanorods arrays grown by chemical bath deposition for ultraviolet photodetectors with high response performance. *J. Alloys Compd.* **2020**, *815*, 152346. [CrossRef]
20. Park, Y.H.; Shin, H.; Noh, S.J.; Kim, Y.; Lee, S.S.; Kim, C.G.; An, K.S.; Park, C.Y. Optical quenching of NiO/Ni coated ZnO nanowires. *Appl. Phys. Lett.* **2007**, *91*, 012102. [CrossRef]
21. Liu, S.; Liao, Q.L.; Zhang, Z.; Zhang, X.K.; Lu, S.N.; Zhou, L.X.; Hong, M.Y.; Kang, Z.; Zhang, Y. Strain modulation on graphene/ZnO nanowire mixed dimensional van der Waals heterostructure for high-performance photosensor. *Nano Res.* **2017**, *10*, 3476–3485. [CrossRef]
22. Li, C.L.; Han, C.; Zhang, Y.B.; Zang, Z.G.; Wang, M.; Tang, X.S.; Du, J.H. Enhanced photoresponse of self-powered perovskite photodetector based on ZnO nanoparticles decorated CsPbBr3 films. *Sol. Energ. Mat. Sol. Cells* **2017**, *172*, 341–346. [CrossRef]
23. Patel, M.; Pataniya, P.M.; Patel, V.; Sumesh, C.K.; Late, D.J. Large area, broadband and highly sensitive photodetector based on ZnO-WS_2/Si heterojunction. *Sol. Energy* **2020**, *206*, 974–982. [CrossRef]
24. Hsiao, Y.J.; Fang, T.H.; Ji, L.W.; Yang, B.Y. Red-shift effect and sensitive responsivity of MoS_2/ZnO flexible photodetectors. *Nanoscale Res. Lett.* **2015**, *10*, 443. [CrossRef]
25. Oh, I.K.; Kim, W.H.; Zeng, L.; Singh, J.; Bae, D.; Mackus, A.J.M.; Song, J.G.; Seo, S.; Shong, B.; Kim, H.; et al. Synthesis of a hybrid nanostructure of ZnO-decorated MoS_2 by atomic layer deposition. *ACS Nano* **2020**, *14*, 1757–1769. [CrossRef] [PubMed]
26. Lan, C.Y.; Li, C.; Wang, S.; Yin, Y.; Guo, H.Y.; Liu, N.S.; Liu, Y. ZnO-WS2 heterostructures for enhanced ultra-violet photodetectors. *RSC Adv.* **2016**, *6*, 67520–67524. [CrossRef]
27. Lv, W.Q.; Liu, J.; He, Y.; You, J.H. Atomic layer deposition of ZnO thin film on surface modified monolayer MoS_2 with enhanced photoresponse. *Ceram. Int.* **2018**, *44*, 23310–23314. [CrossRef]
28. Ma, J.C.; Deng, K.; Zheng, L.; Wu, S.F.; Liu, Z.; Zhou, S.Y.; Sun, D. Experimental progress on layered topological semimetals. *2D Mater.* **2019**, *6*, 032001. [CrossRef]
29. Lai, J.W.; Liu, X.; Ma, J.C.; Wang, Q.S.; Zhang, K.N.; Ren, X.; Liu, Y.N.; Gu, Q.Q.; Zhuo, X.; Lu, W.; et al. Anisotropic broadband photoresponse of layered Type-II weyl semimetal $MoTe_2$. *Adv. Mater.* **2018**, *30*, 1707152. [CrossRef] [PubMed]
30. Zeng, Z.; Wang, D.B.; Wang, J.Z.; Jiao, S.J.; Huang, Y.W.; Zhao, S.X.; Zhang, B.K.; Ma, M.Y.; Gao, S.Y.; Feng, X.; et al. Self-assembly synthesis of the MoS_2/PtCo alloy counter electrodes for high-efficiency and stable low-cost dye-sensitized solar cells. *Nanomaterials* **2020**, *10*, 1725. [CrossRef] [PubMed]
31. Zeng, L.H.; Lin, S.H.; Li, Z.J.; Zhang, Z.X.; Zhang, T.F.; Xie, C.; Mak, C.H.; Chai, Y.; Lau, S.P.; Luo, L.B.; et al. Driven, air-stable, and broadband photodetector based on vertically aligned $PtSe_2$/GaAs heterojunction. *Adv. Funct. Mater.* **2018**, *28*, 1705970. [CrossRef]
32. Lee, Y.F.; Punugupati, S.; Wu, F.; Jin, Z.; Narayan, J.; Schwartz, J. Evidence for topological surface states in epitaxial Bi2Se3 thin film grown by pulsed laser deposition through magneto-transport measurements. *Curr. Opin. Solid State Mater. Sci.* **2014**, *18*, 279–285. [CrossRef]
33. Yu, X.C.; Yu, P.; Wu, D.; Singh, B.; Zeng, Q.S.; Lin, H.; Zhou, W.; Lin, J.H.; Suenaga, K.; Liu, Z.; et al. Atomically thin noble metal dichalcogenide: A broadband mid-infrared semiconductor. *Nat. Commun.* **2018**, *9*, 1–9. [CrossRef]
34. Yin, J.B.; Tan, Z.J.; Hong, H.; Wu, J.X.; Yuan, H.T.; Liu, Y.J.; Chen, C.; Tan, C.W.; Yao, F.R.; Li, T.R.; et al. Ultrafast and highly sensitive infrared photodetectors based on two-dimensional oxyselenide crystals. *Nat. Commun.* **2018**, *9*, 3311. [CrossRef] [PubMed]
35. Wang, D.B.; Jiao, S.J.; Sun, S.J.; Zhao, L.C. Al0 40 in 0.02Ga0.58N Based Metal-semiconductor-metal Photodiodes for Ultraviolet Detection. In Proceedings of the 2012 International Conference on Optoelectronics and Microelectronics, Changchun, China, 23–25 August 2012.
36. Zhang, X.; Zhu, T.S.; Huang, J.W.; Wang, Q.; Cong, X.; Bi, X.Y.; Tang, M.; Zhang, C.R.; Zhou, L.; Zhang, D.Q.; et al. Electric field tuning of interlayer coupling in noncentrosymmetric 3R-MoS_2 with an electric double layer interface. *ACS Appl. Mater. Interfaces* **2020**, *12*, 46900–46907. [CrossRef] [PubMed]
37. Lim, H.; Lee, J.S.; Shin, H.J.; Shin, H.S.; Choi, H.C. Spatially resolved spontaneous reactivity of diazonium salt on edge and basal plane of graphene without surfactant and its doping effect. *Langmuir* **2010**, *26*, 12278–12284. [CrossRef]
38. Ayalakshmi, G.; Saravanan, K. High-performance UV surface photodetector based on plasmonic Ni nanoparticles-decorated hexagonal-faceted ZnO nanorod arrays architecture. *J. Mater. Sci. Mater. Electron.* **2020**, *31*, 5710–5720. [CrossRef]
39. Cheng, H.M.; Hsu, H.C.; Tseng, Y.K.; Lin, L.J.; Hsieh, W.F. Scattering and efficient UV photoluminescence from well-aligned ZnO nanowires epitaxially grown on gan buffer layer. *J. Phys. Chem. B* **2005**, *109*, 8749–8754. [CrossRef]
40. Liu, G.; Kong, L.P.; Hu, Q.Y.; Zhang, S.J. Diffused morphotropic phase boundary in relaxor-$PbTiO_3$ crystals: High piezoelectricity with improved thermal stability. *Appl. Phys. Rev.* **2020**, *7*, 021405. [CrossRef]
41. Zhang, J.; Peng, Z.P.; Soni, A.; Zhao, Y.Y.; Xiong, Y.; Peng, B.; Wang, J.B.; Dresselhaus, M.S.; Xiong, Q.H. Raman spectroscopy of few-quintuple layer topological insulator Bi_2Se_3 nanoplatelets. *Nano Lett.* **2011**, *11*, 2407–2414. [CrossRef] [PubMed]
42. Buchenau, S.; Scheitz, S.; Sethi, A.; Slimak, J.E.; Glier, T.E.; Das, P.K.; Dankwort, T.; Akinsinde, L.; Kienle, L.; Rusydi, A.; et al. Temperature and magnetic field dependent Raman study of electron-phonon interactions in thin films of Bi_2Se_3 and Bi_2Te_3 nanoflakes. *Phys. Rev. B* **2020**, *101*, 245431. [CrossRef]

43. Liu, B.; Liao, Q.L.; Zhang, X.K.; Du, J.L.; Ou, Y.; Xiao, J.K.; Kang, Z.; Zhang, Z.; Zhang, Y. Strain-engineered van der waals interfaces of mixed-dimensional heterostructure arrays. *ACS Nano* **2019**, *13*, 9057–9066. [CrossRef] [PubMed]
44. Lloyd, D.; Liu, X.; Christopher, J.W.; Cantley, L.; Wadehra, A.; Kim, B.L.; Goldberg, B.B.; Swan, A.K.; Bunch, J.S. Band gap engineering with ultralarge biaxial strains in suspended monolayer MoS_2. *Nano Lett.* **2016**, *16*, 5836–5841. [CrossRef]
45. Alwadai, N.; Ajia, I.A.; Janjua, B.; Flemban, T.H.; Mitra, S.; Wehbe, N.; Wei, N.N.; Lopatin, S.; Ooi, B.S.; Roqan, I.S. Catalyst-free vertical ZnO-nanotube array grown on p-GaN for UV-light-emitting devices. *ACS Appl. Mater. Interfaces* **2019**, *11*, 27989–27996. [CrossRef]
46. Djurisic, A.B.; Leung, Y.H. Optical properties of ZnO nanostructures. *Small* **2006**, *2*, 944–961. [CrossRef]
47. Bohle, D.S.; Spina, C.J. The relationship of oxygen binding and peroxide sites and the fluorescent properties of zinc oxide Semiconductor nanocrystals. *J Am. Chem. Soc.* **2007**, *129*, 12380–12381. [CrossRef]
48. Stroyuk, O.L.; Dzhagan, V.M.; Shvalagin, V.V.; Kuchmiy, S.Y. Size-dependent optical properties of colloidal ZnO nanoparticles charged by photoexcitation. *J. Phys. Chem. C* **2010**, *114*, 220–225. [CrossRef]
49. Zhang, H.B.; Zhang, X.J.; Liu, C.; Lee, S.T.; Jie, J.S. High-responsivity, high-detectivity, ultrafast topological insulator bi_2se_3/silicon heterostructure broadband photodetectors. *ACS Nano* **2016**, *10*, 5113–5122. [CrossRef] [PubMed]
50. Liu, C.; Zhang, H.B.; Sun, Z.; Ding, K.; Mao, J.; Shao, Z.B.; Jie, J.S. Topological insulator Bi_2Se_3 nanowire/Si heterostructure photodetectors with ultrahigh responsivity and broadband response. *J. Phys. Chem. C* **2016**, *4*, 5648–5655.
51. Wang, X.; Wu, H.; Wang, G.; Ma, X.; Xu, Y.; Zhang, H.; Jin, L.; Shi, L.; Zou, Y.; Yin, J.; et al. Study of the optoelectronic properties of ultraviolet photodetectors based on Zn-Doped CuGaO2 Nanoplate/ZnO nanowire heterojunctions. *Phys. Status Solidi* **2020**, *257*, 1900684. [CrossRef]
52. Zhu, Y.X.; Liu, K.W.; Wang, X.; Yang, J.L.; Chen, X.; Xie, X.H.; Li, B.H.; Shen, D.Z. Performance improvement of a ZnMgO ultraviolet detector by chemical treatment with hydrogen peroxide. *J. Mater. Chem. C* **2017**, *5*, 7598–7603. [CrossRef]
53. Chen, X.; Liu, K.W.; Wang, X.; Li, B.H.; Zhang, Z.Z.; Xie, X.H.; Shen, D.Z. Performance enhancement of a ZnMgO film UV photodetector by HF solution treatment. *J. Mater. Chem. C* **2017**, *5*, 10645–10651. [CrossRef]
54. Nasiri, N.; Bo, R.; Hung, T.F.; Roy, V.A.L.; Fu, L.; Tricoli, A. Tunable band-selective UV-photodetectors by 3D Self-assembly of heterogeneous nanoparticle networks. *Adv. Funct. Mater.* **2016**, *26*, 7359–7366. [CrossRef]
55. Ismail, R.A.; Al-Naimi, A.; Al-Ani, A.A. Studies on fabrication and characterization of a high-performance Al-doped ZnO/n-Si (111) heterojunction photodetector. *Semicond. Sci. Technol.* **2008**, *23*, 075030. [CrossRef]
56. Ali, A.; Wang, D.B.; Wang, J.Z.; Jiao, S.J.; Guo, F.Y.; Zhang, Y.; Gao, S.Y.; Ni, S.M.; Luan, C.; Wang, D.Z.; et al. ZnO nanorod arrays grown on an AlN buffer layer and their enhanced ultraviolet emission. *CrystEngComm* **2017**, *19*, 6085–6088. [CrossRef]
57. Sun, X.W.; Huang, J.Z.; Wang, J.X.; Xu, Z. A ZnO nanorod inorganic/organic heterostructure light-emitting diode emitting at 342 nm. *Nano Lett.* **2008**, *8*, 1219–1223. [CrossRef] [PubMed]
58. Gupta, A.; Chowdhury, R.K.; Ray, S.K.; Srivastava, S.K. Selective photoresponse of plasmonic silver nanoparticle decorated Bi_2Se_3 nanosheets. *Nanotechnology* **2019**, *30*, 435204. [CrossRef]

Article

Simple and Acid-Free Hydrothermal Synthesis of Bioactive Glass 58SiO$_2$-33CaO-9P$_2$O$_5$ (wt%)

Ta Anh Tuan [1], Elena V. Guseva [1], Nguyen Anh Tien [2], Ho Tan Dat [3] and Bui Xuan Vuong [4,*]

[1] Faculty of Chemical Technologies, Kazan National Research Technological University, 420015 Tatarstan, Russia; taanhtuan84pt@hpu2.edu.vn (T.A.T.); leylaha@mail.ru (E.V.G.)
[2] Faculty of Chemistry, Ho Chi Minh City University of Education, Ho Chi Minh City 700000, Vietnam; tienna@hcmue.edu.vn
[3] Chu Van An Secondary School, Ho Chi Minh City 700000, Vietnam; tandatsphoahoc@gmail.com
[4] Faculty of Pedagogy in Natural Sciences, Sai Gon University, Ho Chi Minh City 700000, Vietnam
* Correspondence: bxvuong@sgu.edu.vn

Abstract: The paper focuses on the acid-free hydrothermal process for the synthesis of bioactive glass. The new method avoids the use of harmful acid catalysts, which are usually used in the sol-gel process. On the other hand, the processing time was reduced compared with the sol-gel method. A well-known ternary bioactive glass 58SiO$_2$-33CaO-9P$_2$O$_5$ (wt%), which has been widely synthesized through the sol-gel method, was selected to apply to this new process. Thermal behavior, textural property, phase composition, morphology, and ionic exchange were investigated by thermal analysis, N$_2$ adsorption/desorption, XRD, FTIR, SEM, and inductively coupled plasma optical emission spectrometry (ICP-OES) analysis. The bioactivity and biocompatibility of synthetic bioactive glass were evaluated by in vitro experiments with a simulated body fluid (SBF) solution and cell culture medium. The obtained results confirmed that the acid-free hydrothermal process is one of the ideal methods for preparing ternary bioactive glass.

Keywords: bioactive glass; acid-free hydrothermal; bioactivity; hydroxyapatite; cell viability

Citation: Anh Tuan, T.; V. Guseva, E.; Anh Tien, N.; Tan Dat, H.; Vuong, B.X. Simple and Acid-Free Hydrothermal Synthesis of Bioactive Glass 58SiO$_2$-33CaO-9P$_2$O$_5$ (wt%). *Crystals* **2021**, *11*, 283. https://doi.org/10.3390/cryst11030283

Academic Editors: Cölfen Helmut and Zarkov Aleksej

Received: 3 February 2021
Accepted: 8 March 2021
Published: 12 March 2021

Publisher's Note: MDPI stays neutral with regard to jurisdictional claims in published maps and institutional affiliations.

Copyright: © 2021 by the authors. Licensee MDPI, Basel, Switzerland. This article is an open access article distributed under the terms and conditions of the Creative Commons Attribution (CC BY) license (https://creativecommons.org/licenses/by/4.0/).

1. Introduction

Bioactive glasses (BGs) have been applied as bone fillers within the clinic for the past fifty years owing to their osteoconductivity and osteoinductivity [1]. These materials possess a special ability to attach to the bone tissue through the formation of a mineral hydroxyapatite layer when soaked in a biological solution. In this way, broken and defective bones are repaired and filled [2]. Since the primary discovery of bioactive glass (45SiO$_2$-24.5CaO-24.5Na$_2$O-6P$_2$O$_5$, wt%; the commercial name is Bioglass® or Novamin), many glass systems have been studied and synthesized using two main methods: melting and sol-gel [3,4]. In particular, the sol-gel method shows outstanding advantages compared to the melting method. The sol-gel processes are often applied to fabricate bioactive glasses at lower temperatures, preventing the loss of the ultimate product due to the evaporation of P$_2$O$_5$. Specifically, this method can synthesize bioactive glasses on a nanoscale with a high value of the specific surface area and porous structures, which improves the bioactivity of synthetic materials [5,6]. Nevertheless, most sol-gel processes have used harmful strong acids and bases as catalysts for the hydrolysis of precursors [7–17]. This affects the manufacturer's health when synthesizing the glass systems as well as the possibility of toxic acid residues in the final synthetic product. Therefore, finding acid-free synthesis methods can be seen as an urgent need to create friendly products that can be used as artificial bone materials or as additives in healthcare products. Recently, some scientists have reported on the synthesis of bioactive glasses by the green chemical process. Following this trend, new synthesis processes are being established to cut back or eliminate the utilization or generation of harmful substances [18,19]. A quaternary bioactive glass 75%SiO$_2$-16%CaO-5%Na$_2$O-4%P$_2$O$_5$ (mol%) has been synthesized by the acid-free sol-gel process [20]. The authors rapidly added the precursors TEOS and TEP in an exceedingly large volume of water, under strong stirring of 1100 rpm. In this way, the alkoxides hydrolyzed completely to create a transparent sol after 5 h. In our previous studies, binary bioactive glasses 70Si-30Ca (mol%) with or without Zn doping, were made by the acid-free hydrothermal process [21,22]. The mixtures of suitable precursors without catalytic acids were heated in a Teflon-lined stainless steel autoclave at 150 °C in an electric oven and kept there for 1 day. The resulting product, in gel form, was dried and heat-treated at 700 °C for 3 h. The obtained glasses were amorphous materials and showed interesting bioactivities. For this paper, we applied the acid-free hydrothermal method to synthesize a well-known sol-gel bioactive glass 58SiO$_2$-33CaO-9P$_2$O$_5$ (wt%). Moreover, the advantage of the selected Na-free ternary system SiO$_2$-CaO-P$_2$O$_5$ avoids the negative effect of a sodium element. The release of sodium ions can cause a rapid pH increase, which can result in a cytotoxic effect [23–25]. By changing some conditions, like hot temperatures and the water/TEOS molar ratio, compared to previous studies [21,22], we emphasize that the acid-free hydrothermal process can be completely applied to synthesizing ternary bioactive glass 58SiO$_2$-33CaO-9P$_2$O$_5$ (wt%). The synthetic glass was characterized and examined for its bioactivity and biocompatibility.

2. Materials and Methods

2.1. Acid-Free Hydrothermal Synthesis

The acid-free hydrothermal process was used to prepare the bioactive glass 58SiO$_2$-33CaO-9P$_2$O$_5$ (wt%). The composition of the glass system was selected as in previous studies, where the material system was synthesized by the sol-gel method [14–17]. A brief description, a mixture containing 10.42 g of TEOS (tetraethyl orthosilicate, Sigma-Aldrich, $\geq$ 99.0%, Pcode: 102068011), 1.21 g of TEP (triethyl phosphate, Merck, 100%, CAS-No:78-40-0), 7.09 g of Ca(NO$_3$)$_2$.4H$_2$O (calcium nitrate tetrahydrate, Merck, 100%, CAS-No-13477-34-4), and 54 g of H$_2$O was stirred for 30 min. The H$_2$O/TEOS molar ratio was surveyed and selected at 60. After being mixed together, the mixture was placed in a stainless steel autoclave lined with a Teflon core. The hydrothermal synthesis reactor was programmed at 160 °C for 24 h. The gel-producing product was dried at 100 °C for 24 h to form the gel powder. From the thermal analysis data, the bioactive glass was obtained by sintering the dried powder at around 700 °C, with a heating rate of 10 °C/min, and kept there for 3 h.

2.2. In Vitro Experiment in SBF Fluid

An in vitro test is critical to confirm the bioactivity of synthetic biomaterials before in vivo tests in the animal body. The in vitro test was proposed by Kokubo and Takadama through the immersion of the material in the simulated body fluid (SBF) and widely applied for bioactivity evaluation [26]. The SBF synthetic solution had concentrations of inorganic ions almost similar to the blood of a human body (Table 1). It was synthesized by dissolving the appropriate chemical agents comprising MgCl$_2$.6H$_2$O, CaCl$_2$, K$_2$HPO$_4$.3H$_2$O, NaHCO$_3$, KCl, NaCl, and C$_4$H$_{11}$NO$_3$ in distilled water at a body temperature of 37 °C and a pH of 7.4. The powdered glasses were immersed in the SBF solution at 37 °C for 1, 3, and 5 days, with a stirring speed maintained at 60 rpm. At the end of each soaking stage, the powdered samples were refined, dried, and used for chemical–physical characterization. The remaining solutions were used for ionic measurements.

Table 1. Ionic concentration of the simulated body fluid (SBF) solution (mmol/L).

Composition	Na$^+$	K$^+$	Ca^{2+}	Mg^{2+}	Cl$^-$	HCO$_3^-$	HPO$_4^{2-}$
SBF	142.0	5.0	2.5	1.5	148.0	4.2	1.0
Plasma	142.0	5.0	2.5	1.5	103.0	27.0	1.0

2.3. In Vitro Assay Within the Cellular Medium

The cell culture medium was Dulbecco's Modified Eagle's Medium (DMEM; Merck, Product Code D9785) containing 10% fetal bovine serum (FBS), 100 µg mL^{-1} of penicillin, 10 µg mL^{-1} of streptomycin, and 15 mM of HEPES (4-(2-hydroxyethyl)-1-piperazineethanesulfonic acid). The L-929 fibroblast line was cultured in DMEM at a temperature of 37 °C in a humid incubator (5% CO_2, 95% humidity) for 24 h. The ratio of glass powder/medium was selected as 0.1 g mL^{-1} according to the ISO standard 10993-12:2004. The various dilutions were obtained from the extract of the cellular medium, named as 20%, 40%, 60%, and 100% (without dilution). The fibroblast cells were exposed to the extracts for 24 h. The cellular viabilities on the bioactive glass were evaluated by the MTT (3-(4,5-Dimethylthiazol-2-yl)-2,5-Diphenyltetrazolium Bromide) method consistent with the previous study [27].

2.4. Characterization

The thermal behavior of the as-sintering bioactive glass was obtained by employing a Thermogravimetry–Differential Scanning Calorimetry (TG–DSC; Labsys Evo Setaram, Thermal Analysis Labs Ltd., Fredericton, NB, Canada). The fine glass powder was put in a platinum crucible and then heated up from 30 to 1000 °C at 10 °C/min in dried air. The textural properties were obtained by using N_2 adsorption/desorption on a micromeritics porosimeter (Quantachrome Instruments, Boynton Beach, FL, USA). The specific surface area was achieved by using the Brunauer–Emmett–Teller (BET) technique (Micrometrics, Georgia, USA). The pore size and pore volume were calculated from the isotherm desorption curve based on the Barrett–Joyner–Halanda (BJH) method. The phase characteristics of the powder samples were identified by X-ray diffraction (XRD; D8-Advance, Bruker, Billerica, MA, USA) with Cu-K$_\alpha$ radiation (λ = 1.5406 Å). The measurements were performed within the range of 5–80° (2θ), with a step of 0.02°. The XRD identification was performed by the X-Pert High Score Plus software. The chemical bonding groups were determined by a Fourier-Transform Infrared Spectroscopy (FTIR, Bruker Equinox 55, Bruker, Billerica, MA, USA). The spectral scan was carried out in the range of 400–4000 cm^{-1}, with a resolution of 2 cm^{-1}. Scanning Electron Microscopy (SEM) combined with Energy Dispersive X-ray Spectroscopy (EDX) (S-4800, Hitachi, Tokyo, Japan) was employed to identify the morphology and elemental composition of the powder samples. The ionic exchange in the SBF solution during the in vitro experiment was verified by using inductively coupled plasma optical emission spectrometry (ICP-OES, ICP 2060, Agilent, California, USA).

3. Results and Discussion

3.1. Thermal Behavior

The thermogravimetry (TG) and the differential scanning calorimetry (DSC) curves of the dried sample are presented in Figure 1. The TG curve shows three mass losses in the temperature ranges of 28–210, 210–405, and 405–670 °C. The primary mass loss corresponding to the endothermic peak at 129.4 °C on the DSC curve, is assigned to the physically adsorbed water removal [21,28]. The second one, with an exothermic peak at 298.2 °C on the DSC curve, is characteristic of the chemically adsorbed water release [29]. The last one, with an endothermic peak centered at 521.1 °C on the DSC branch, was due to the thermal decomposition of the NO_3^- groups used as oxide precursors [30,31]. An exothermic peak at 923.6 °C without the mass loss is attributed to the formation of wollastonite $CaSiO_3$ compound, according to our previous studies [21,22]. From temperatures above 670 °C, no mass loss was observed. Therefore, the suitable temperature for glass sintering is chosen to be at around 700 °C to eliminate the H_2O and NO_3^- components within the sample.

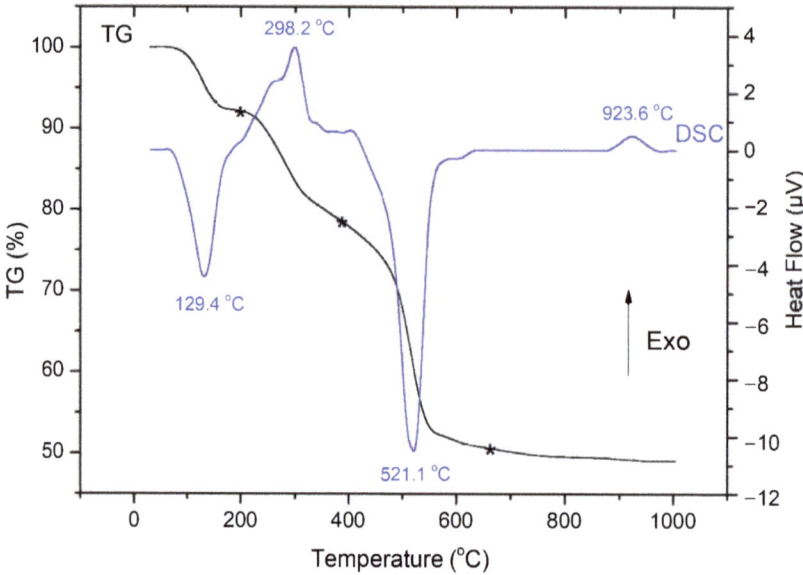

Figure 1. Thermogravimetry–Differential Scanning Calorimetry (TG–DSC) curves of the as-sintering bioactive glass sample.

3.2. Textural Analysis

The N_2 adsorption/desorption isotherm and pore size distribution of the bioactive glass powder are shown in Figure 2. The synthetic bioactive glass exhibited the type IV isotherm, which is suitable for mesoporous material based on the IUPAC classification of adsorption isotherms [32]. The BJH pore size distribution achieved from the desorption branch shows a relatively wide range and a single-type distribution with pore sizes of 8 to 90 nm, concentrated at a mean diameter (MD) of 21.2 nm. The measured values of the specific surface area (SSA) and pore volume (PV) are 104.7 m^2/g and 0.54 cm^3/g, respectively. In this study, the synthetic bioactive glass presents interesting values of SSA, PV, and MD compared to previous papers, as shown in Table 2 [15–17].

Table 2. Textural properties of synthetic bioactive glass.

Reference	Specific Surface Area (m^2/g)	Pore Volume (cm^3/g)	Mean Diameter (nm)	Synthesis Method
[15]	82	0.201	10	Sol-Gel
[16]	99.1	-	-	Sol-Gel
[17]	126.54	0.447	6.55	Sol-Gel
[This Study]	104.7	0.54	21.2	Hydrothermal

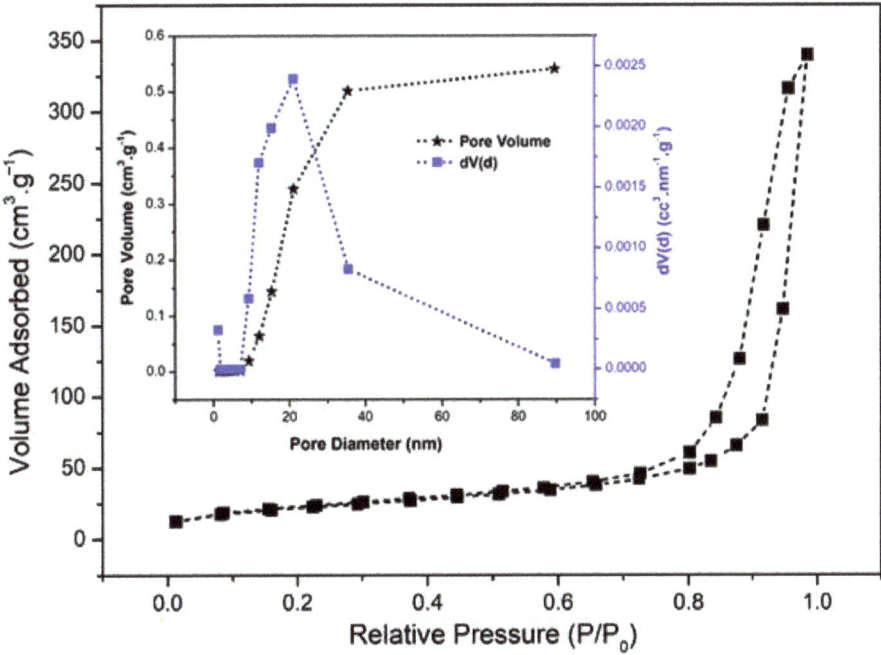

Figure 2. Nitrogen adsorption/desorption isotherm and pore size distribution of bioactive glass.

3.3. Bioactivity Evaluation

3.3.1. XRD Analysis

Figure 3 shows the XRD diagrams of glass samples before and after the in vitro experiments in the SBF solution. The XRD pattern of the synthetic bioactive glass exhibits a wide diffraction halo centered at about 23° (2θ). This is a typical characteristic of an amorphous material, confirming the successful synthesis of ternary bioactive glass $58SiO_2$-$33CaO$-$9P_2O_5$ (wt%) by the free-acid hydrothermal method. After the primary day in the SBF solution, the bioactive glass showed two well-defined peaks at 2θ = 26° (002) and 32° (211), which are attributed to the crystalline hydroxyapatite (HA) phase (well-matched with the JCPDS file no. 90432). As the soaking time increased, these two peaks became sharper and higher in intensity. After 5 days, most of the representative peaks of the HA phase appeared clearly, demonstrating the bioactivity of the synthetic bioactive glass in this study.

3.3.2. FTIR Analysis

Figure 4 represents the FTIR spectra of the bioactive glass before and after the in vitro experiment in the SBF. The spectrum of the synthetic glass represents the most characteristic bands of the silica network. The band at 1104 cm^{-1} is attributed to the Si-O-Si asymmetric stretching vibration (asym) while the band at 803 cm^{-1} corresponds to the Si-O-Si symmetric stretching vibration (sym) [33]. The band observed at 470 cm^{-1} is characteristic of the Si-O-Si bending vibration [34]. The weak band at 977 cm^{-1} is ascribed to the Si-O non-bridging oxygen (NBO) stretching mode [34]. Only a weak band at 554 cm^{-1}, related to the stretching mode of PO_4^{3-} [35,36], was observed because of the low content of P_2O_5 in the synthetic bioactive glass. After the in vitro experiment in the SBF for 1, 3, and 5 days, the spectral feature of the glass was modified, thanks to the chemical interactions between the glass samples and the physiological medium. The spectral band at 1104 cm^{-1} (Si-O-Si asym) was shifted to 1030 cm^{-1}. The band at 803 cm^{-1} (Si-O-Si sym) was moved to

869 cm^{-1}. The band at 470 cm^{-1} (Si-O-Si ben) was displaced to 460 cm^{-1}. The band at 977 cm^{-1} (Si-O-NBO) disappeared. The Si-O-Si shift and Si-O disappearance are associated with the glassy-network dissolution, and then the re-polymerization of the -Si(OH)$_4$ groups to make the SiO$_2$-rich surface layer [32–36]. Typically, two well-defined bands at 564 and 600 cm^{-1} are revealed. They are attributed to the stretching modes of the PO$_4^{3-}$ groups in the hydroxyapatite crystals [37,38]. The FTIR results associated with the XRD analysis emphasize the bioactivity of bioactive glass synthesized by the free-acid hydrothermal method.

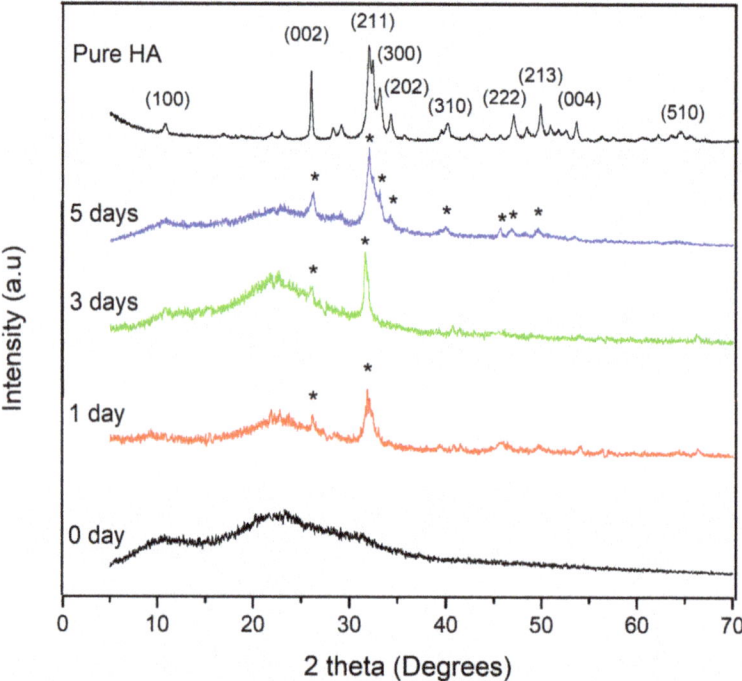

Figure 3. XRD diagrams of bioactive glass before and after the in vitro experiment in the SBF.

3.3.3. SEM–EDX Analysis

Figure 5 shows the SEM micrographs, including EDX analyses, of the glass samples before and after immersion in the SBF fluid for 5 days. The SEM image of the synthetic glass shows agglomerates consisting of intertwined tiny particles, forming the three-dimensional mesoporous structure of the synthetic material. The EDX analysis gives the Si/Ca/P molar ratio of 7.52:4.41:1, which is quite similar to the theoretical ratio in synthetic glass (60SiO$_2$-36CaO-4P$_2$O$_5$ mol%; Si/Ca/P = 7.5/4.5/1). After 5 days of immersion in the SBF solution, the surface of the bioactive glass was replaced and completely recovered by a uniform, scaled crystal layer. The EDX analysis of the glass after 5 days in SBF shows a decrease in Si content because of the degradation of the glassy network, and a rise in Ca and P amounts due to the formation of HA phase. The calculated Ca/P molar ratio for 5 days is 1.72, similar to that of pure apatite [30,31]. The SEM observation and EDX analysis confirmed the appearance of a new apatite layer on the surface of the bioactive glass.

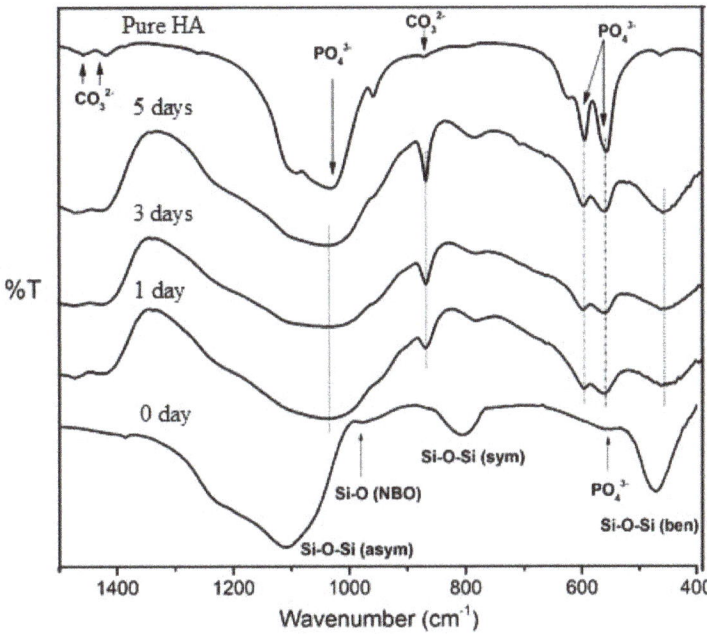

Figure 4. FTIR spectra of bioactive glass before and after the in vitro experiment in the SBF.

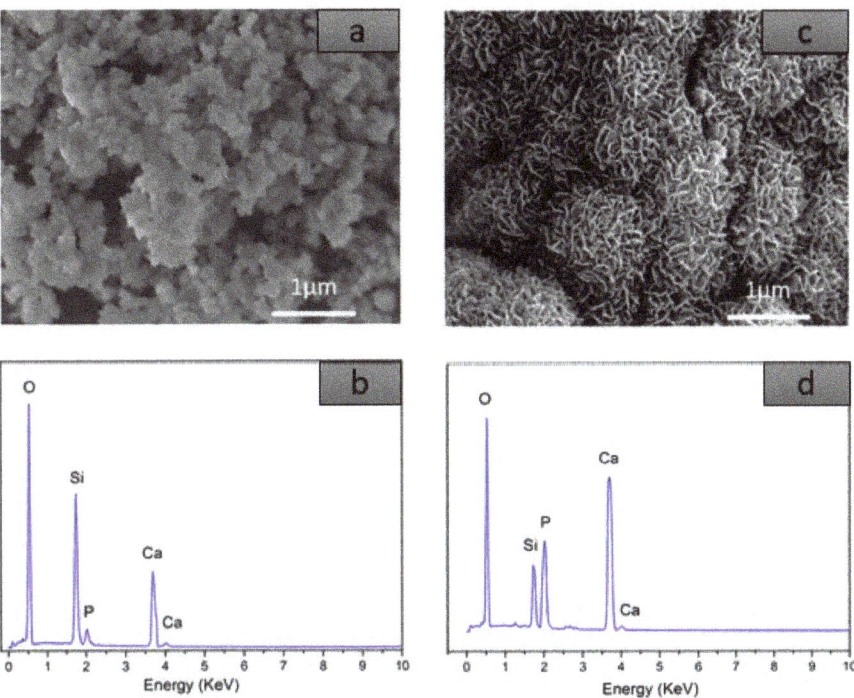

Figure 5. SEM–Energy Dispersive X-ray Spectroscopy (EDX) analyses of the bioactive glass: (**a,b**) before and (**c,d**) after 5 days of immersion.

3.3.4. ICP-OES Analysis

The physical–chemical interactions of the bioactive glass with the physiological environment led to ionic changes in the SBF solution as presented in the Figure 6. The elemental concentrations of Si, Ca, and P in the initial SBF solution were 0 ppm, 100.2 ppm, and 31.4 ppm, respectively. The Si concentration rapidly increased at the beginning time of soaking, and then moderately increased after 3 days. The concentration of Si reached the saturation value after 5 days of immersion. According to the previous studies, a rising in Si concentration is explained by the dissolution of the glassy network through the release of silicic acid $Si(OH)_4$, while the saturation process corresponds to the re-polymerization of the above acids to create a SiO_2 silica layer [33–36]. On the other hand, the concentration of Ca increases at the beginning time of soaking, probably due to the quick exchange of Ca^{2+} out of the glassy network and H^+ in the physiological fluid [37–39]. Thereafter, the concentration of Ca strongly decreased after 3 days and reached saturation at 5 days. The decrease of the Ca concentration is expounded by its consumption to make the mineral HA layer on the surface of the bioactive glass [37–39]. By contrast, no increase in P concentration was observed after the in vitro experiment. This can be explained by the low content of P_2O_5 in the synthetic glass and also the rapid consumption of Ca and P for the formation of the apatite mineral layer. This result completely fits with the XRD analysis, where the HA layer was determined after just 1 day of soaking in the SBF solution.

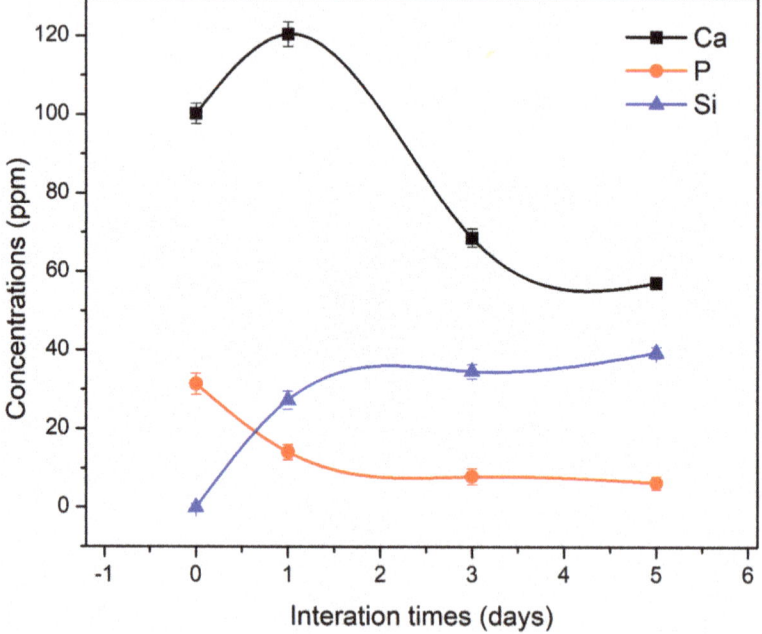

Figure 6. Ionic exchanges between the bioactive glass and the SBF solution.

3.4. Biocompatibility Evaluation

The cell viabilities of the L-929 fibroblast cells directly in contact with the bioactive glass powder for 24 h are presented in Figure 7. The cell viability without contact with the bioactive glass was selected as the control (100%) [27]. Following the standard ISO 10993-5 (Biological evaluation of medical devices—Part 5: Test for cytotoxicity, in vitro methods 2009), the cell viability was calculated as a percentage relative to the control, set as 100%. In the case where the average of cell viability was less than 70%, the material was cytotoxic. The obtained results show that the cell viabilities were 124, 116, 96, 94% for 20%, 40%, 60%, and 100% extracts, respectively. The 20% extract showed the highest value of cell viability, while the 60% and 100% extracts presented a small difference. Therefore, the bioactive glass $58SiO_2$-$33CaO$-$9P_2O_5$ (wt%) synthesized by the acid-free hydrothermal method presented good biocompatibility in the cellular medium even within the high extracts. The value of cell viability for the bioactive glass in this study is equivalent to those for previous glass systems, such as 45S5 melting Bioglass®, 77S sol-gel glass, and 58S sol-gel glass [40].

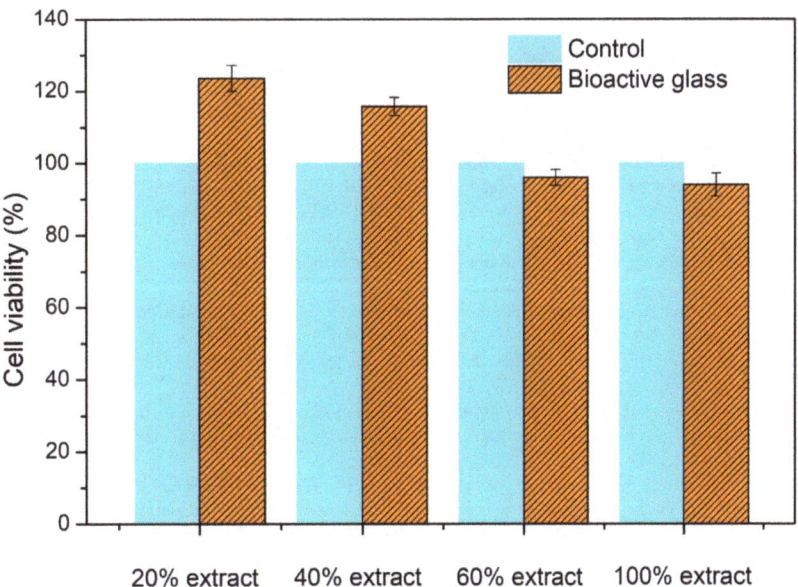

Figure 7. Cell viabilities of synthetic bioactive glass.

4. Conclusions

This study confirmed that the acid-free hydrothermal method is suitable for the synthesis of ternary bioactive glass $58SiO_2$-$33CaO$-$9P_2O_5$ (wt%). The obtained bioactive glass has amorphous and mesoporous structures. The bioactivity of the synthetic glass was proved through the rapid formation of the hydroxyapatite mineral layer on the surface of glass samples after an in vitro experiment in SBF. In addition, the in vitro experiment in the cellular environment demonstrated the good biocompatibility of the synthetic bioactive glass. The acid-free hydrothermal synthesis can be considered a new method to synthesize bioactive glass with properties similar to previous studies. The bioactive glass synthesized by this process, which does not use catalytic acids, can be used as artificial bone materials, and also can be used in food technology, such as additions to toothpaste or cosmetics.

Author Contributions: T.A.T.—Methodology, analysis, writing; E.V.G.—Writing, editing; N.A.T.—Writing, editing; H.T.D.—Methodology, analysis; B.X.V.—Writing, editing, analysis. All authors have read and agreed to the published version of the manuscript.

Funding: This research was funded by Sai Gon University, grant number TĐ 2020-2.

Institutional Review Board Statement: Not applicable.

Informed Consent Statement: Not applicable.

Data Availability Statement: Not applicable.

Conflicts of Interest: The authors declare no conflict of interest.

References

1. Rahaman, M.N.; Day, D.E.; Bal, B.S.; Fu, Q.; Jung, S.B.; Bonewald, L.F.; Tomsia, A.P. Bioactive glass in tissue engineering. *Acta Biomater.* **2011**, *7*, 2355–2373. [CrossRef] [PubMed]
2. Hench, L.L. The story of Bioglass®. *J. Mater. Sci. Mater. Med.* **2006**, *17*, 967–978. [CrossRef] [PubMed]
3. Julian, R.J. Review of bioactive glass: From Hench to hybrids. *Acta Biomater.* **2013**, *9*, 4457–4486.
4. Baino, F.; Novajra, G.; Pacheco, M.P.; Boccaccini, A.R.; Brovarone, C.V. Bioactive glasses: Special applications outside the skeletal system. *J. Non. Cryst. Solids.* **2016**, *432*, 15–30. [CrossRef]
5. Boccaccini, A.R.; Erol, M.; Stark, W. Polymer/bioactive glass nanocomposites for biomedical applications: A review. *Compos. Sci. Technol.* **2010**, *70*, 1764–1776. [CrossRef]
6. Zheng, K.; Boccaccini, A.R. Sol-gel processing of bioactive glass nanoparticles: A review. *Adv. Colloid Interface Sci.* **2017**, *249*, 363–373. [CrossRef]
7. Sharifianjazi, F.; Parvin, N.; Tahriri, M.J. Synthesis and characteristics of sol-gel bioactive SiO_2-P_2O_5-CaO-Ag_2O glasses. *Non Cryst. Solids.* **2017**, *476*, 108–113. [CrossRef]
8. Vulpoi, A.; Gruian, C.; Vanea, E.; Baia, L.; Simon, S.; Steinhoff, H.J.; Goller, G.; Simon, V. Silver effect on the structure of SiO_2-CaO-P_2O_5 ternary system. *Mater. Sci. Eng. C* **2012**, *32*, 178–183. [CrossRef]
9. Balamurugan, A.; Balossier, G.; Kannan, S.; Michel, J.; Rebelo, A.H.S.; Ferreira, J.M.F. Development and in vitro characterization of sol–gel derived CaO–P_2O_5–SiO_2–ZnO bioglass. *Acta Biomater.* **2007**, *3*, 255–262. [CrossRef]
10. Salman, S.; Salama, S.; Abo-Mosallam, H. The role of strontium and potassium on crystallization and bioactivity of Na_2O–CaO–P_2O_5–SiO_2 glasses. *Ceram. Int.* **2012**, *38*, 55–63. [CrossRef]
11. El-Kheshen, A.A.; Khaliafaa, F.A.; Saad, E.A.; Elwana, R.L. Effect of $Al2O3$ addition on bioactivity, thermal and mechanical properties of some bioactive glasses. *Ceram. Int.* **2008**, *34*, 1667–1673. [CrossRef]
12. Baino, F.; Fiume, E.; Miola, M.; Leone, F.; Onida, B.; Verné, E. Fe-doped bioactive glass-derived scaffolds produced by sol-gel foaming. *Mater. Lett.* **2019**, *235*, 207–211. [CrossRef]
13. Bari, A.; Bloise, N.; Fiorilli, S.; Novajra, G.; Vallet-Regí, M.; Bruni, G.; Torres-Pardo, A.; González-Calbet, J.M.; Visai, L.; Vitale-Brovarone, C. Copper-containing mesoporous bioactive glass nanoparticles as multifunctional agent for bone regeneration. *Acta Biomater.* **2017**, *55*, 493–504. [CrossRef] [PubMed]
14. Bini, M.; Grandi, S.; Capsoni, D.; Mustarelli, P.; Saino, E.; Visai, L.J. SiO_2– P_2O_5– CaO glasses and glass-ceramics with and without ZnO: Relationships among composition, microstructure, and bioactivity. *Phys. Chem. C* **2009**, *113*, 8821–8828. [CrossRef]
15. Bejarano, J.; Caviedes, P.; Palzal, H. Sol–gel synthesis and in vitro bioactivity of copper and zinc-doped silicate bioactive glasses and glass-ceramics. *Biomed. Mater.* **2015**, *10*, 025001. [CrossRef]
16. Bui, X.V.; Dang, T.H. Bioactive glass 58S prepared using an innovation sol-gel process. *Process. App. Ceram.* **2019**, *13*, 98–103. [CrossRef]
17. Sepulveda, P.; Jones, J.R.; Hench, L.L. Characterization of melt-derived 45S5 and sol-gel–derived 58S bioactive glasses. *J. Biomed. Mater. Res.* **2001**, *58*, 734–740. [CrossRef]
18. Sheldon, R.A.; Arends, I.W.C.E.; Hanefeld, U. *Green Chemistry and Catalysis*; Wiley VCH: Weinheim, Germany, 2007.
19. Clark, J.H.; Macquarrie, D.J. *Green Chemistry and Technology*; Abingdon: Nashville, TN, USA, 2008.
20. Ben-Arfa, B.A.E.; Fernandes, H.R.; Salvado, I.M.M.; Ferreira, J.M.F.; Pullar, R.C. Effects of catalysts on polymerization and microstructure of sol-gel derived bioglasses. *J. Am. Ceram. Soc.* **2018**, *101*, 2831–2839. [CrossRef]
21. Hoa, B.T.; Hoa, H.T.T.; Tien, N.A.; Khang, N.H.D.; Guseva, E.V.; Tuan, T.A.; Vuong, B.X. Green synthesis of bioactive glass 70SiO2-30CaO by hydrothermal method. *Mater. Lett.* **2020**, *274*, 128032. [CrossRef]
22. Tuan, T.A.; Guseva, E.V.; Phuc, L.H.; Hien, N.Q.; Long, N.V.; Vuong, B.X. Acid-free Hydrothermal Process for Synthesis of Bioactive Glasses 70SiO2–(30-x)CaO–xZnO (x = 1, 3, 5 mol.%). *Proceed.* **2020**, *62*, 1–12.
23. Fernandes, H.R.; Gaddam, A.; Rebelo, A.; Brazete, D.; Stan, G.E.; Ferreira, J.M.F. Bioactive glasses and glass-ceramics for healthcare applications in bone regeneration and tissue engineering. *Materials* **2018**, *11*, 2530. [CrossRef]
24. Wallace, K.E.; Hill, R.G.; Pembroke, J.T.; Brown, C.J.; Hatton, P.V. Influence of sodium oxide content on bioactive glass properties. *J. Mater. Sci. Mater. Med.* **1999**, *10*, 697–701. [CrossRef]
25. Kansal, I.; Reddy, A.; Muñoz, F.; Choi, S.; Kim, H.; Tulyaganov, D.U.; Ferreira, J.M.F. Structure, biodegradation behavior and cytotoxicity of alkali-containing alkaline-earth phosphosilicate glasses. *Mater. Sci. Eng. C* **2014**, *44*, 159–165. [CrossRef]
26. Kokubo, T.; Takadama, H. How useful is SBF in predicting in vivo bone bioactivity? *Biomaterials* **2006**, *27*, 2907–2915. [CrossRef] [PubMed]
27. Mosmann, T.J. Rapid colorimetric assay for cellular growth and survival: Application to proliferation and cytotoxicity assays. *Immunol. Meth.* **1983**, *65*, 55–63. [CrossRef]
28. Saboori, A.; Rabiee, M.; Moztarzadeh, F.; Sheikhi, M.; Tahriri, M.; Karimi, M. Synthesis, characterization and in vitro bioactivity of sol-gel-derived SiO2–CaO–P2O5–MgO bioglass. *Mater. Sci. Eng. C* **2009**, *29*, 335–340. [CrossRef]
29. El-Kady, A.M.; Ali, A.F. Fabrication and characterization of ZnO modified bioactive glass nanoparticles. *Ceram. Int.* **2012**, *38*, 1195–1204. [CrossRef]
30. Delben, J.R.J.; Pereira, K.; Oliveira, S.L.; Alencar, L.D.S.; Hernandes, A.C.; Delben, A.A.S.T. Bioactive glass prepared by sol–gel emulsion. *J. Non-Cryst. Solids.* **2013**, *361*, 119–123. [CrossRef]

31. Román, J.; Padilla, S.; Vallet-Regí, M. Sol− gel glasses as precursors of bioactive glass ceramics. *Chem. Mater.* **2003**, *15*, 798–806. [CrossRef]
32. Thommes, M.; Kaneko, K.; Neimark, A.V.; Olivier, J.P.; Rodriguez-Reinoso, F.; Rouquerol, J.; Sing, K.S.W. Physisorption of gases, with special reference to the evaluation of surface area and pore size distribution (IUPAC Technical Report). *Pure Appl. Chem.* **2015**, *87*, 1051–1069. [CrossRef]
33. Aina, V.; Malavasi, G.; Pla, A.F.; Munaron, L.; Morterra, C. Zinc-containing bioactive glasses: Surface reactivity and behaviour towards endothelial cells. *Acta Biomater.* **2009**, *5*, 1211–1222. [CrossRef] [PubMed]
34. Kim, I.Y.; Kawachi, G.; Kikuta, K.; Cho, S.B.; Kamitakahara, M.; Ohtsuki, C.J. Preparation of bioactive spherical particles in the CaO–SiO2 system through sol–gel processing under coexistence of poly (ethylene glycol). *Eur. Ceram. Soc.* **2008**, *28*, 1595–1602. [CrossRef]
35. Innocenzi, P.J. Infrared spectroscopy of sol–gel derived silica-based films: A spectra-microstructure overview. *Non-Cryst. Solids* **2003**, *316*, 309–319. [CrossRef]
36. Ding, J.; Chen, Y.; Chen, W.; Hu, L.; Boulon, G. Effect of P_2O_5 addition on the structural and spectroscopic properties of sodium aluminosilicate glass. *Chin. Opt. Lett.* **2012**, *10*, 071602. [CrossRef]
37. Chen, X.F.; Lei, B.; Wang, Y.J.; Zhao, N.J. Morphological control and in vitro bioactivity of nanoscale bioactive glasses. *Non-Cryst. Solids.* **2009**, *355*, 791–796. [CrossRef]
38. Hong, Z.; Liu, A.; Chen, L.; Chen, X.; Jing, X.J. Preparation of bioactive glass ceramic nanoparticles by combination of sol–gel and coprecipitation method. *Non-Cryst. Solids* **2009**, *355*, 368–372. [CrossRef]
39. Hoppe, A.; Güldal, N.S.; Boccaccini, A.R. A review of the biological response to ionic dissolution products from bioactive glasses and glass-ceramics. *Biomaterials* **2011**, *32*, 2757–2774. [CrossRef]
40. Silver, I.A.; Deas, J.; Hska, M.E. Interactions of bioactive glasses with osteoblasts in vitro: Effects of 45S5 Bioglass®, and 58S and 77S bioactive glasses on metabolism, intracellular ion concentrations and cell viability. *Biomaterials* **2001**, *22*, 175–185. [CrossRef]

Article

Synthesis and Characterization of Potent and Safe Ciprofloxacin-Loaded Ag/TiO$_2$/CS Nanohybrid against Mastitis Causing *E. coli*

Naheed Zafar [1], Bushra Uzair [1,*], Muhammad Bilal Khan Niazi [2], Ghufrana Samin [3], Asma Bano [4], Nazia Jamil [5], Waqar-Un-Nisa [6], Shamaila Sajjad [7] and Farid Menaa [8]

1. Department of Biological Sciences, International Islamic University, Islamabad 44000, Pakistan; naheedzafar12@gmail.com
2. School of Chemical and Materials Engineering, National University of Sciences and Technology, Islamabad 44000, Pakistan; bilalniazi2@gmail.com
3. Department of Chemistry, Faisalabad Campus, University of Engineering and Technology Lahore, Faisalabad 38000, Pakistan; g.samin@uet.edu.pk
4. Department of Microbiology, University of Haripur, Haripur 22620, Pakistan; asma_baano@yahoo.com
5. Institute of Microbiology & Molecular Genetics, Punjab University, Lahore 54000, Pakistan; nazia.mmg@pu.edu.pk
6. Centre for Interdisciplinary Research in Basic and Applied Sciences, International Islamic University, Islamabad 44000, Pakistan; waqarunnisa@iiu.edu.pk
7. Department of Physics, International Islamic University, Islamabad 44000, Pakistan; shamaila.sajjad@iiu.edu.pk
8. Department of Nanomedicine and Advanced Technologies, California Innovations Corporation, San Diego, CA 92037, USA; dr.fmenaa@gmail.com
* Correspondence: bushra.uzair@iiu.edu.pk

Citation: Zafar, N.; Uzair, B.; Niazi, M.B.K.; Samin, G.; Bano, A.; Jamil, N.; Waqar-Un-Nisa; Sajjad, S.; Menaa, F. Synthesis and Characterization of Potent and Safe Ciprofloxacin-Loaded Ag/TiO$_2$/CS Nanohybrid against Mastitis Causing *E. coli*. *Crystals* **2021**, *11*, 319. https://doi.org/10.3390/cryst11030319

Academic Editor: Jaime Gómez Morales

Received: 22 February 2021
Accepted: 12 March 2021
Published: 23 March 2021

Publisher's Note: MDPI stays neutral with regard to jurisdictional claims in published maps and institutional affiliations.

Copyright: © 2021 by the authors. Licensee MDPI, Basel, Switzerland. This article is an open access article distributed under the terms and conditions of the Creative Commons Attribution (CC BY) license (https://creativecommons.org/licenses/by/4.0/).

Abstract: To improve the efficacy of existing classes of antibiotics (ciprofloxacin), allow dose reduction, and minimize related toxicity, this study was executed because new target-oriented livestock antimicrobials are greatly needed to battle infections caused by multidrug-resistant (MDR) strains. The present study aims to green synthesize a biocompatible nanohybrid of ciprofloxacin (CIP)-Ag/TiO$_2$/chitosan (CS). Silver and titanium nanoparticles were green synthesized using *Moringa concanensis* leaves extract. The incorporation of silver (Ag) nanoparticles onto the surface of titanium oxide nanoparticles (TiO$_2$NPs) was done by the wet chemical impregnation method, while the encapsulation of chitosan (CS) around Ag/TiO$_2$ conjugated with ciprofloxacin (CIP) was done by the ionic gelation method. The synthesized nanohybrid (CIP-Ag/TiO$_2$/CS) was characterized using standard techniques. The antibacterial potential, killing kinetics, cytotoxicity, drug release profile, and minimum inhibitory concentration (MIC) were determined. Field emission scanning electron microscopy (FESEM) and transmission electron microscopy (TEM) revealed spherical agglomerated nanoparticles (NPs) of Ag/TiO$_2$ with particle sizes of 47–75 nm, and those of the CIP-Ag/TiO$_2$/CS nanohybrid were in range of 20–80 nm. X-ray diffractometer (XRD) patterns of the hetero system transmitted diffraction peaks of anatase phase of TiO$_2$ and centered cubic metallic Ag crystals. Fourier Transform Infrared spectroscopy (FTIR) confirmed the Ti-O-Ag linkage in the nanohybrid. The zeta potential of CIP-Ag/TiO$_2$/CS nanohybrid was found (67.45 ± 1.8 mV), suggesting stable nanodispersion. The MIC of CIP-Ag/TiO$_2$/CS was 0.0512 µg/mL, which is much lower than the reference value recorded by the global CLSI system (Clinical Laboratory Standards Institute). The CIP-Ag/TiO$_2$/CS nanohybrid was found to be effective against mastitis causing MDR *E. coli*; killing kinetics showed an excellent reduction of *E. coli* cells at 6 h of treatment. Flow cytometry further confirmed antibacterial potential by computing 67.87% late apoptosis feature at 6 h of treatment; antibiotic release kinetic revealed a sustained release of CIP. FESEM and TEM confirmed the structural damages in MDR *E. coli* (multidrug-resistant *Escherichia coli*). The CIP-Ag/TiO$_2$/CS nanohybrid was found to be biocompatible, as more than 93.08% of bovine mammary gland epithelial cells remained viable. The results provide the biological backing for the development of nanohybrid antibiotics at a lower MIC value to treat infectious diseases of cattle and improve the efficacy of existing classes of antibiotics by conjugation with nanoparticles.

Keywords: ciprofloxacin; TiO$_2$/Ag/CS nanohybrid; ionic gelation; mastitis; MDR *E. coli*

1. Introduction

The use of antibiotics in the treatment of animal infection and the enhancement of animal health are the main driving forces behind the development of antimicrobial resistance in animal husbandry [1,2]. Drug resistance has been attributed to the over use or misuse of antibiotics in the cattle to treat mastitis, whereas MDR (multidrug-resistant) strains induced mastitis in cattle and are the major cause of economic loss in the dairy industry, and it is also more likely due to irrational use of antibiotics in the dairy industry [3]. In addition, traditional antibiotics have been less effective in the treatment of infectious diseases in dairy cattle. In this context, there is an urgent need for the production of biocompatible antibacterial formulations that could regulate bacterial growth by means of improved and efficient mechanisms [4]. In New York State, a study reported about 82% of milk fed to calves contained residues of antibiotics [5]. Subsequent research studied the effect of the intake of antimicrobial-containing raw milk and identified an increased incidence of antimicrobial resistance in *E. coli* strains of milk/dairy origin [3]. The rise of the antimicrobial resistance is considered a "ticking time bomb" although the WHO (world health organization) has called for a global reduction in veterinary antibiotics to phase out the use of medicinally essential antibiotics by the food industry because of scientific evidence that antimicrobial resistance strains were found in live stocks [6,7]. Ciprofloxacin is a fluoroquinolone class antimicrobial that is widely used for a wide range of infections, including those caused by *E. coli* but resistance is reported against it [8,9].

Nanotechnology is a possible response to antimicrobial resistance, which could promote creativity and create a new generation of antibiotic therapies for potential medicines. Sustaining existing antibiotic activity through novel formulation using nanotechnologies can increase the therapeutic longevity of anti-infection action of existing antibiotics that are no longer effective against MDR strains. There is credible evidence of the effective use of nanotechnologies as antimicrobials [10]. Metal oxide nanoparticles (NPs) as antimicrobial agents demonstrated relatively high efficiency to combat MDR strains. Metal-based nanomaterials, such as silver nanoparticles (AgNPs) and titanium, have attracted immense attention because of their excellent efficiency against MDR bacteria owing to their electronic, optical, and catalytic properties [11]. After the generation of Ag$^+$ and Ti$^+$ ions by the oxidation of Ag NPs, TiO$_2$ eventually induced reactive oxygen radicals (ROS) inside the MDR *E. coli*. The free radicals ($^{\bullet}$O$_2^-$, $^{\bullet}$OH, and H$_2$O$_2$) are generated due to the doping of Ag on the surface of TiO$_2$ NPs, which also shifting absorbance of TiO$_2$ NPs from the UV region to visible region of (200–800 nm), enhancing optical properties ultimately and producing ROS species ($^{\bullet}$O$_2^{--}$, $^{\bullet}$OH, and H$_2$O$_2$) that cause immediate damage to the MDR *E. coli* and leakage of cellular constituents, leading to the death of the bacterial cell. In addition, it is relatively simple to synthetize AgNPs, and the mass production of metal oxide-based Agnanohybrids is cost-effective [12]. The antimicrobial activity of TiO$_2$ was first reported by Matsunaga and colleagues in 1988. Under near-UV light illumination, microbial cells could be destroyed by contact with a TiO$_2$–Pt [13]. Nanomaterials based on chitosan have gained considerable interest in the biomedical field due to their unusual biodegradability, biocompatibility, and protection against heart diseases, as chitosan could decrease cholesterol absorption and antimicrobial properties [14]. This powerful biopolymer is capable of enhancing the stability of AgNPs and Ag-based nanohybrids [15]. Chitosan-based composite nonmaterials have been reported as highly effective antibacterial agents and found wide applications including in drug delivery, tissue engineering, wound healing, and antibacterial activity [16]. There are many ways to develop new drug delivery systems using liposomes, polymer–drug conjugates, lipid-based nanoparticles, and copolymericcells, which could improve the drug delivery [17].

Recently, the plant-mediated green synthesis of silver nanoparticles has grown into a unique and novel field of nanotechnology. It has gained importance because of its eco-friendly and cost efficiency with lower toxicity as compared to chemical methods [18]. The reduction properties of plant secondary metabolites are responsible for the enhanced ability of plant extracts to produce nanoparticles with enhanced properties [19]. *M. concanensisnimmo* is a medicinal herb with versatile use in pharmaceutical products, antibacterial agent, food source, and water-purifying agent, and it is used for the treatment of variety of diseases including paralysis, menstrual pain, high blood pressure, skin tumors, liver and kidney disease, as well as to treat inflammation of joints, indicating its immense importance in health care industry [20,21].

Considering the urgent need for new clinical interventions to control the global issue of antimicrobial resistance (AMR), a unique ciprofloxacin loaded-TiO_2/Ag/CS nanohybrid was synthesized by the green approach using *M. concanensis* leaves extract. AgNPs and TiO_2NPs were prepared by *M. concanensis* leaves extract, as a reducing and stabilizing agent. The incorporation of AgNPs onto the surface of TiO_2 nanomaterials was performed by the wet chemical impregnation technique. The chitosan (CS) encapsulation of TiO_2/Ag composite followed after coupling with ciprofloxacin (CIP) by the ionic gelation method. Physical characterizations were performed by the routine state-of-the-art techniques of microscopy and spectroscopy. A ciprofloxacin loaded-TiO_2/Ag/CS nanohybrid was tested for its efficacy against MDR strains of *E. coli*-causing mastitis in the cattle by various standard antimicrobial assays indicating the strong antibacterial potential of synthesized nanohybrids. Eventually, we assessed the possible sustained release of the ciprofloxacin from the hybrid material by a drug release kinetics study. Cytotoxicity was evaluated by MTT (3-(4, 5-dimethylthiazol-2-yl)-2, 5-diphenyl tetrazolium bromide) assay for veterinary application using bovine mammary gland cell lines.

2. Experimental Section/Materials & Methods

2.1. Chemicals and Reagents

Ethanol (C_2H_5OH), acetic acid (CH_3COOH), titanium tetraisoproxide (TTP), silver nitrate ($AgNO_3$), and chitosan medium molecular weight (Product number 448877, 75–85% deacetylation, 200–800 cP viscosity of 1% w/v in 1% v/v acetic acid), pentasodiumtripolyphosphate (TPP), and glacial acetic acid were all purchased from Sigma-Aldrich (Sigma-Aldrich Co., St Louis, MO, USA). Propidium iodide (PI), Annexin V and in the LIVE/DEAD®BacLight™ Bacterial Viability Kit (L7012) (ThermoFisher Scientific, UK), MTT (3-(4, 5-dimethylthiazol-2-yl)-2, 5-diphenyl tetrazolium bromide) assay kit (Cell Proliferation Kit MTT Sigma-Aldrich), Dulbecco's Modified Eagle Medium (DMEM) (Sigma-Aldrich). The Gram staining chemicals, Nutrient Broth (NB), Nutrient Agar (NA), Muller–Hinton Agar (MHA),MacConkey Agar, Brain–Heart Infusion (BHI) broth, andantibiotic discs were all obtained from Oxoid (Thermo Fisher, England, UK), API 20 E system(France, Biomerieux).

2.2. Isolation and Identification of MDR E. coli Strains

MDR *E. coli* strains (N = 87) were isolated from milk samples collected by National Veterinary Laboratories of Pakistan. The California mastitis test was used for the detection of mastitis, and *E. coli* strains were identified using MacConkey agar (Oxoid, UK) as selective medium for selective isolation of *E. coli* from the milk samples. A standard API20E test panel was used for the biochemical-based identification of *E. coli* [22].

2.3. Antibiotic Sensitivity Testing

The disc diffusion assay was used to perform antibiotics sensitivity profiling of isolated *E. coli* strains [22]. According to the antibiotics commonly prescribed for mastitis treatment, the following antibiotic discs were used: Ceftazidime (CAZ) 30 µg, Cefazolin (KZ) 30 µg, Cefoxitin (FOX) 30 µg, Imipenem (IPM) 10 µg, Ceftriaxone (CRO) 30µg, Ampicillin (AMP) 10 µg, Ciprofloxacin (CIP) 5 µg, Meropenem (MEM) 10 µg, Augmentin(AMC) 20 µg,

Gentamicin (CN) 10 µg, Doxycycline (DO) 30 µg, Norfloxacin (NOR) 30 µg, Fosfomycin (FOS) 10 µg, Tetracycline (TE) 30 µg, and Trimethoprim/Sulfamethoxazole (SXT) 30 µg. Discs were placed with the help of a disc dispenser on the MHA (Muller–Hinton Agar) plate and incubated for 24 h at 37 °C. The respective zone of inhibition (ZI) was measured by using a vernier caliper, and the results were compared with those available in the CLSI (Clinical Laboratory Standards Institute) guidelines [23]. A disc dispenser was used to place each disc on pre-inoculated Mueller–Hinton (MH) agar plates, which were incubated for 24 h at 37 °C. Subsequently, ZIs (zones of inhibition) were recorded, and the data were interpreted following CLSI guidelines [24].

2.4. Plant Material Collection and Extract Preparation

M. concanensis was used for the green synthesis of silver and titanium nanoparticles; the collected plant was authenticated by Professor Mushtaq, Department of Botany, QAU Islamabad *M. Concanensis* (Hoon Pakistan). The extract of leaves was used for the synthesis of NPs. The leaves were washed and dried. They were used as a reducing agent for the amalgamation of NPs. A total of 50 g dried leaves was dissolved into 500 mL distilled water (dH$_2$O) in a beaker and kept on a hot plate at 55 °C for 10 h. The extract has been cooled down at room temperature (RT) and purified for further use.

2.5. Green Synthesis of TiO$_2$ and Ag Nanoparticles

For the preparation of TiO$_2$ NPs and AgNPs as shown in Scheme 1, 5 mL of the plant extract was added either to 100 mL titanium tetraisoproxide (TTP) solution (0.4 M) or to 1 mM solution of silver nitrate (AgNO$_3$) under mild magnetic stirring at 28 °C. A change in color of the resultant solution from colorless to brown confirmed the reduction of Ag+ to Ag° moreover in case of TiO$_2$ color turned to creamy white. Each mixture was centrifuged three times at 12,000 rpm to extract unreacted ions for 10 min. The respective final products were dried at 60 °C, ground, and calcined at 500 °C for 3 h in a muffle furnace to get pure NPs [25].

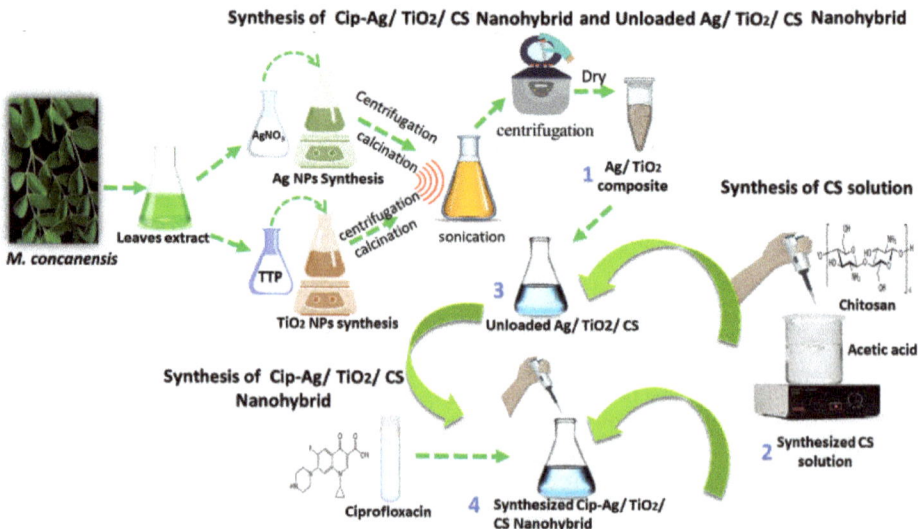

Scheme 1. Green synthesis of nanoformulations using *M. concanensis* leaves extract.

2.6. Formation of TiO$_2$/Ag Nanocomposite

Ag/TiO$_2$ nanocomposite was prepared by the wet chemical impregnation method [26]. For that purpose, AgNPs and TiO$_2$ NPswere mixed in equal amount (0.01 g) into 10 mL of ethanol, and the mixture was homogenized by sonication for 3 h. The precipitate was obtained after 3–5 times washing and centrifugation at 6000 rpm at 25 °C. The final mixture was dried at 60 °C temperature to obtain the final product Ag/TiO$_2$, as shown in Scheme 1.

2.7. Preparation of CIP-Ag/TiO$_2$/CS Nanohybrid and Unloaded Ag/TiO$_2$/CS

A CIP-Ag/TiO$_2$/CS nanohybrid was obtained by mixing 4 mL of TPP solution containing TiO$_2$/Ag nanocomposite 0.01 g with CIP 0.01 g, and continuous magnetic stirring at 28 °C for 2 h. After complete homogenization of the mixture, the flask was subjected to ultrasonication at 35 Hz for 30 min to scatter the solution particles and agglomeration. Subsequently, the mixture was centrifuged at 14,000 rpm for 30 min at 4 °C [27], the supernatant was discarded, and the pellet was resuspended in ddH$_2$O until use. The same procedure was followed for preparing an unloaded Ag/TiO$_2$/CS nanohybrid; only CIP was eliminated in the reacting mixture, as shown in Scheme 1.

2.8. Physical Characterization of the Green Synthesized Nanoformulations

An X-ray diffractometer (D/MAX 2550, Rigaku Ltd., Tokyo, Japan) was used to determine the crystalline nature and the crystal phase composition of the prepared nanoformulations [28]. The diffraction was observed in the range of 10–80° (2θ) wide-angle XRD using (Cu Kα1 radiation, λ = 1.5406 Å) at 40 kV and 100 mA. The Scherrer equation was used to estimate the crystalline dimension. A Fourier Transform Infrared (FTIR) spectrometer (FT-IR Spectrum 100 spectrometer, PerkinElmer, Waltham, MA, USA)was used to determine the functional groups of unknown elements at 500–5000 cm^{-1} [29]. TEM (JEM 1010, JEOL EU, Nieuw-Vennep, The Netherlands) was used to analyze the size and shape of the nanohybrid at 200 keV, while the selected area electron diffraction (SAED) was used for phase analysis. TEM at 175 nm of scale was also used to study morphological changes in bacterial cells after treatment with CIP-loaded nanohybrid. Field emission scanning electron microscopy (MIRA3TESCAN Inc., Warrendale, PA, USA) photographs at 500 nm scale and EDX (energy-dispersive X-ray) spectra of nanoformulations were used to analyze the size and shape of the nanohybrid. EDX was used to reveal the composition of the elements in the synthesized nanoformulations. The zeta potential of the nanoformulations was measured at room temperature through a Malvern zeta seizer. The electrical charge on the diffused aqueous layer that formed on the surface of the NPs when submerged in water was used to improve the stability of the dosage and the shelf life and to minimize the time and expense of the formulation. The zeta potential and particle size of all the synthesized NPs were measured at room temperature. To perform the following characterizations i.e., XRD, FTIR, TEM, FESEM, and EDX analysis, the dry powder form of all the synthesized nanoformulationswas used directly. To avoid agglomerations of nanoformulations and ensure accuracy, results were obtained after sonication for 40 min. Zeta potential analysis was executed by making suspensions of all the synthesized nanoformulations in the aqueous media.

2.9. Determination of Encapsulation Efficiency

Efficiency of drug encapsulation was determined in accordance with Liu's protocol [30]. A nanophotometer (Implen) was used to measure the quantity of encapsulated CIP into a CIP-TiO$_2$/Ag/CS nanohybrid. The nanohybridwas isolated from the drug-free nanohybrid (i.e., TiO$_2$/Ag/CS) by centrifugation at 10,000 rpm for 25 min, and the supernatant was quantified by spectrophotometry at a wavelength of λ_{max} (295 nm). The encapsulation efficiency (EE) of CIP into CIP-TiO$_2$/Ag/CS nanohybrid was determined using the following equation:

EE = [Loaded CIP into nanohybrid/Total amount of nanohybrid (free + loaded)] × 100

2.10. Antibacterial Activity of Green Synthesized Nanoformulations

The green synthesized nanoformulations, i.e., CIP-Ag/TiO$_2$/CS nanohybrid, unloaded Ag/TiO$_2$/CS nanohybrid, CIP, AgNPs, TiO$_2$NPs, Ag/TiO$_2$ nanocomposite, and

CS were evaluated by the disc diffusion method [24]. The petri plates that contained Muller–Hinton Agar standard media were seeded with MDR strains of *E. coli*. Various concentrations of nanoformulation were screened for anti *E. coli* activity. Sterile filter paper disks were loaded with nanoformulations and placed aseptically on the seeded MH agar plate gently, DMSO was considered as negative control, and unloaded Ag/TiO$_2$/CS nanohybrid discs served as (internal control). The zones of inhibition (ZIs) were recorded after 24 h incubation at 37 °C, the experiment was repeated in triplicate, and SD (standard deviation) was calculated.

2.11. MIC Determination of CIP-Ag/TiO$_2$/CS Nanohybrid

The minimum inhibitory concentration (MIC) of the synthetic antibiotic CIP was determined by using broth microdilution in Nutrient Broth (NB), according to the E-strip method [31]. Briefly, MDR *E. coli* strains suspensions were prepared by comparing the turbidity of the overnight inoculated culture with the 0.5 McFarland standards. Then, each suspension showing turbidity was swabbed with the help of a sterile cotton swab on an MH agar plate, and an E-strip was placed aseptically on the surface of agar. The plates were incubated aerobically for 24 h at 37 °C, after which the data were recorded and analyzed by using a global CLSI-based system [32]. The MIC of the CIP-Ag/TiO$_2$/CS nanohybrid was performed following the broth microdilution method in NB, with a few modifications. Briefly, MDR *E. coli* suspensions showing turbidity according to the 0.5 McFarland standard were added in the test tubes containing different concentrations (0–10 µg/mL) of the CIP-Ag/TiO$_2$/CS nanohybrid before incubation for 18 h at 37 °C. Subsequently, the dilution containing the lowest concentration capable of inhibiting the bacterial growth by displaying visible broth media is called the MIC value. Positive (inoculated broth only) and negative controls (NB only) were included in the experiment.

2.12. Kinetics of Antibacterial Effects of CIP-Ag/TiO$_2$/CS Nanohybrid

In the preliminary experiments, the MIC for each nanoformulation was determined, and exponentially growing *E. coli* cells were used to inoculate a series of tubes containing 2 mL of nutrient broth (Oxoid) (A600 nm 0.05) and a set concentrations of nanoformulations (CIP-Ag/TiO$_2$/CS nanohybrid, unloaded Ag/TiO$_2$/CS nanohybrid, CIP, Ag NPs, TiO$_2$ NPs, Ag/TiO$_2$ nanocomposite, and CS NPs). The time-kill experiments were performed in triplicate and as described by Uzair [33]. Glass flasks that had 50 mL of fresh nutrient broth were inoculated with exponentially growing cells of *E. coli* (A600 nm). Inoculation was carried out following the addition of nanoformulation at the respective concentration of the MIC. The flasks were incubated for 24 h at 30 °C, and the OD (optical density) was calculated at different time points (i.e., 0, 2, 4, 6, 8, 12, and 24 h). The experiment was run in triplicate to authenticate results.

2.13. FESEM Analysis of CIP-Ag/TiO$_2$/CS Nanohybrid

The morphological effect on *E. coli* cells after interaction with CIP-Ag/TiO$_2$/CS nanohybrid at respective MIC concentration was investigated at various time of intervals (i.e., 0, 6, and 12 h) by FE-SEM (Field Emission Scanning Electron Microscopy). Concisely, a tiny drop (10 L) of treated and untreated (control) MDR *E. coli* cells (1×10^4) were put on a glass slide, and slides were placed in 2% glutaraldehyde (GA) and paraformaldehyde HEPES (4-(2-hydroxyethyl)-1-piperazineethanesulfonic acid) buffer (30 mM) for one hour at 37 °C. Then, an increasing concentration of alcohol in water was used for dehydration of the cells, and the slides were kept on water–alcohol solutions for 10 min each in each container having alcohol/water gradient solution. Eventually, slides were washed using ter butyl alcohol for one min, and dried slides were subjected to gold sputter coating for 60 s from three directions before their observation under the FE-SEM microscope (at 2 µm of scale).

2.14. TEM Analysis of CIP-Ag/TiO$_2$/CS Nanohybrid

Ultrastructure changes on *E. coli* log phase cells (1×10^4) were observed under TEM before (control) and after treatment with CIP-Ag/TiO$_2$/CS nanohybrid at respective MIC concentrations for 0, 6, and 12 h. Briefly, 10 µL of treated or untreated bacterial cells were put on a glass slide, which was then fixed using GA 2.5% at 4 °C. After an overnight incubation in GA, the slides were washed with phosphate buffer solution (PBS) 1X. *E. coli* cells were eventually examined under TEM (at 300 nm of field zooming edge)

2.15. Live/Dead Assessment of CIP-Ag/TiO$_2$/CS Nanohybrid-Treated Bacteria

To study Live/Dead *E. coli*, cells treated and untreated with CIP-Ag/TiO$_2$/CS nanohybrid were determined by flow cytometer [34], using AnnexinV and Propidium Iodide (PI) double staining according to the manufacturer's protocol (CUS Ever bright Inc., Suzhou, China). Briefly, a Log-Phase bacterial suspension (1×10^8 CFU/mL) was centrifuged at 3000 rpm for 20 min, and the pellet was resuspended in 50 µL PBS 1X (pH 7.4). Then, *E. coli* cells were exposed to 0.5 µg/µL of CIP-Ag/TiO$_2$/CS nanohybrid in the NB medium at 37 °C and incubated for 6 h. After treatment, *E. coli* cells were centrifuged at 1000 rpm for 15 min, the supernatant was discarded, and the pellet was resuspended in 500 µL of PBS 1X (pH 7.4). The cells were carefully labeled with 10 µL of Annexin V binding buffer and 5 µL of Annexin V-FITC (Fluorescein isothiocyanate) followed by 5 µL of PI stain and incubated in the dark for 20 min at RT, and the resulting stained cells were diluted with 200 µL PBS 1X (pH 7.4) and 400 µL of Annexin binding buffer. After staining, cells were analyzed with FACS (fluorescence-activated single cell sorting) can flow cytometer (Becman Coulter Cytomics FC500).

2.16. Ex Vivo Drug Release Kinetics of CIP-Ag/TiO$_2$/CS Nanohybrid

A Franz diffusion cell was used for ex vivo drug release kinetics studies of the CIP-Ag/TiO$_2$/CS nanohybrid [35]. The skin of a healthy rabbit was taken and fixed between the compartments of the diffusion cell with the donor compartment. Ciprofloxacin (550 µg/mL) alone as control and the CIP-Ag/TiO$_2$/CS nanohybrid solution (at MIC 0.0512 µg/mL present in 3 mL of PBS of pH 7.4) was placed in the donor compartment. The receiver compartment contained 13 mL of PBS of pH 7.4, and the contents were stirred using a magnetic stirrer. The whole assembly was kept at 37 ± 0.5 °C to maintain normal temperature for skin. Test samples of the nanohybrids (3 mL) were taken at preset time points up to 24 h and replenished with PBS pH 7.4. Samples were purified through a syringe filter, and the drug content of the samples was measured using a UV/visible spectrophotometer (Shimadzu Pharmspec1700, Shimadzu Inst. Japan) at 278 nm. The total sum of the drug released was determined using the formula below.

$$\text{Percent of drug release } (\%) = \frac{\text{Absorbance of sample (nm)}}{\text{Absorbance of control (nm)}} \times 100$$

2.17. Ex Vivo Cytotoxicity Study

To evaluate the cytotoxicity of TiO$_2$/Ag nanocomposites, CSNPs, pure CIP, Ag/TiO$_2$/CS nanohybrids, and CIP-Ag/TiO$_2$/CS nanohybrids, MTT assay [36] was used on bovine mammary gland epithelial cells (BMGE). The cell lines of BMGE tissues were grown on Dulbecco's modified Eagle medium (DMEM) in 96-well plates and then (1×10^5 cells) distributed into wells before incubation for 24 h in the CO$_2$ incubator thermo stated at 37 °C. The viable BMGE cells (1×10^5) were treated with nanoformulations at increasing concentrations (0.2, 0.1, and 0.02 µg/mL) incubated at 37 °C for 24 h, keeping Celecoxib as PC (Positive control) and PBS as controls of the study. Following incubation, 100 µL of fresh DMEM was thoroughly mixed with 10 µL of MTT solution prepared in PBS 1X to replace the existing DMEM [36]. The 96-well plates were incubated again for 4 h. Eventually, 0.1 mL of DMSO solution was used to dissolve the formazan crystals in the wells, and the OD of the wells containing the MTT formazan, which was used as an internal control and BMGE cells treated with nanoformulations at various concentrations were computed at the reference wavelengths of 570 nm and 620 nm, respectively. The percentage of viability was observed using the given standard equation:

$$\text{Percent cell viability} = \frac{(\text{Test 570 nm} - 620 \text{ nm})}{(\text{Control 570 nm} - 620 \text{ nm})} \times 100.$$

2.18. Hemolysis Assay

The method for the hemolysis assay of Ag NPs, TiO$_2$ NPs, Ag/TiO$_2$ nanocomposite, unloaded Ag/TiO$_2$/CSNPs, CIP alone, CS NPs, and CIP-Ag/TiO$_2$/CS nanohybrids with PBS (negative control) and Triton 100x (positive control) as controls was performed according to the literature [37]. From the healthy volunteer, blood (3 m) was taken with the help of a sterile syringe in the vacutainers with proper consent of the person, who was informed briefly that the provided blood samples will be exclusively processed for research purposes only. RBCs (red blood cells 1.5 mL) were incubated at 37 °C after treatment with all the synthesized nanoformulations (50 µL) for 6 h, and then spinning was done at 1500 rpm to separate the RBCs from blood. After that, 100 µL of the supernatant of all samples was transferred to a 96-well plate. The absorbance values of the supernatant were taken at 570 nm by using a micro plate reader. The percentage of hemolysis of RBCs was determined by the following equation:

$$\text{Percent (\%) cell viability} = \frac{(\text{Absorbance of Sample} - \text{Negative control})}{(\text{Absorbance of Positive Control} - \text{Negative control})} \times 100.$$

2.19. Ethical Approval and Informed Consent

Pakistan research council regulations were followed in strict accordance with the recommendations in the Guide for the Care and Use of Laboratory Animals of the National Institute of Health (NIH), Islamabad. Experimental protocols (Reg #22-FBAS/PHDBT/F-14) were approved by Institutional Bioethical Committee of International Islamic University, Islamabad.

2.20. Statistics

Mean (SD) was deliberated from independent triplicated experiments. Prevalence was obtained in cross-tabulations and expressed as percentage (%). All statistical analyses were performed using GraphPad Prism 8.1 (GraphPad Software, San Diego, CA, USA).

3. Results and Discussion

3.1. Isolation, Identification, and MIC Determination

E. coli strains were isolated from milk samples producing lactose fermenting pink colonies on MacConkey agar. *E. coli* produced pink colored, smooth, and round colonies [3], which showed the same morphological character of the colony. Under microscope, *E. coli* showed rods, and retained counterstain safranin pink rods of *E. coli* showed a Gram-negative reaction. All the biochemical results were in accordance with previous literature [38]. The rapid identification test for *E. coli*-generated number from API 20E was 5144552. In the present investigation, the disc diffusion method was employed to determine the resistance spectra against multidrug-resistant *E. coli* and the ATCC (8739) strain of *E. coli* as a control organism. The disc diffusion method is the primary tool to classify the strains as an MDR and ESBL (Extended spectrum beta-lactamase) producing *E. coli* strains [32,39]. The results showed that *E. coli* was found highly resistant to the current range of antibiotics, raising concern to find the solution for the emerging resistant Gram-negative rods of mastitis-suffered cattle, as shown in the Figure 1. The minimum inhibitory concentrations were determined by employing the E-strip method and broth microdilution method. Ciprofloxacin by the E-strip method showed more than a 4 µg/mL MIC value by comparing the reference of the global CLSI-based system that is far beyond the recommended dose for *E. coli* [32,39]. Broth microdilution disclosed the MIC of CIP-Ag/TiO$_2$/CS nanohybrid as 0.0512 µg/mL against MDR *E. coli*, which proposed it as an efficient alternative therapeutic agent.

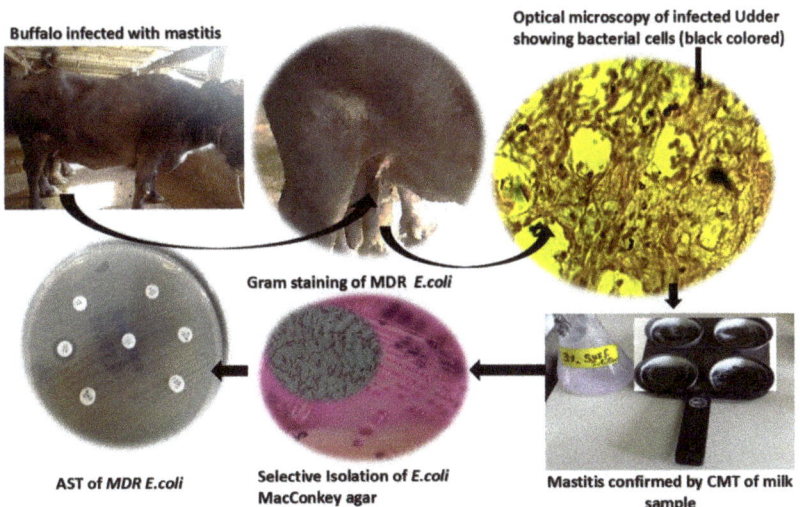

Figure 1. Isolation of multidrug-resistant (MDR) *E. coli* as causative agent of mastitis-induced udder infections in cattle.

3.2. Physical Characterization of the Green Synthetized Nanoformulations

3.2.1. FESEM and TEM Depicted Spherical Morphology and Confirmed the Nano Size of CIP-TiO$_2$/Ag/CS Hybrid

Texture and morphological analyses of greenly synthesized TiO$_2$NPs, AgNPs, Ag/TiO$_2$ nanocomposites, and CIP-Ag/TiO$_2$/CS nanohybrids were determined by FESEM (Figure 2). The the nanostructures of the TiO$_2$NPs, AgNPs, and Ag/TiO$_2$ nanocomposites showed uniform round spherical morphology [40], as illustrated in FESEM micrographs represented by Figure 2A–C, respectively. The particle size ranges of TiO$_2$NPs, AgNPs, Ag/TiO$_2$ nanocomposite CSNPs, and CIP-Ag/TiO$_2$/CS nanohybrids were 25–55 nm, 22–40 nm, 19–35 nm, and 19–75 nm, respectively. This also tentatively confirms the particle sizes observed from XRD analyses. Agglomerated and spherical AgNPs were well dispersed throughout the surface of TiO$_2$ (Figure 2B). It can be remarkably observed from the Ag/TiO$_2$ nanocomposite that AgNPs are incorporated on the surface of TiO$_2$ (Figure 2C). It is noted that there is no distinction between the TiO$_2$NPs and AgNPs in Ag/TiO$_2$ nanocomposites. Eventually, it was clearly seen that the biopolymer CS anchored the whole surface of spherical Ag/TiO$_2$ nanocomposites (Figure 2D). After CS grafting, it was perceived that AgNPs remained segregated onto the TiO$_2$ surface (Figure 2D).

Meanwhile, EDX analysis was done to investigate the elemental distribution of the four nanostructures (Figure 2). Figure 2A displayed the Ti and O signals supporting the TiO$_2$ NP synthesis. Figure 2C revealed the peaks corresponding to the Ti, O, and Ag in Ag/TiO$_2$nanocomposites. It was clear from the signal that a 1.2 wt% nominal content of Ag was closed to its stoichiometric value of 2.0 wt% solution of AgNPs exploited for the fabrication of Ag/TiO$_2$ nanocomposites. The signal of C in Ag/TiO$_2$nanocomposites can be ascribed to the carbon substrate/grid. No additional peaks were observed, which indicated the purity level of the synthesized nanoformulations. Figure 2B showed an elemental profile of AgNPs, which determined the sharp signal of the Ag element. Finally, a synthesized CIP-Ag/TiO$_2$/CS nanohybrid elemental spectrum is shown in Figure 2D, which displayed the peak signals corresponding to TiO$_2$, Ag, TiO$_2$/Ag, and CS, while other peaks or modifications in signal intensity may be attributed to the incorporation of CIP.

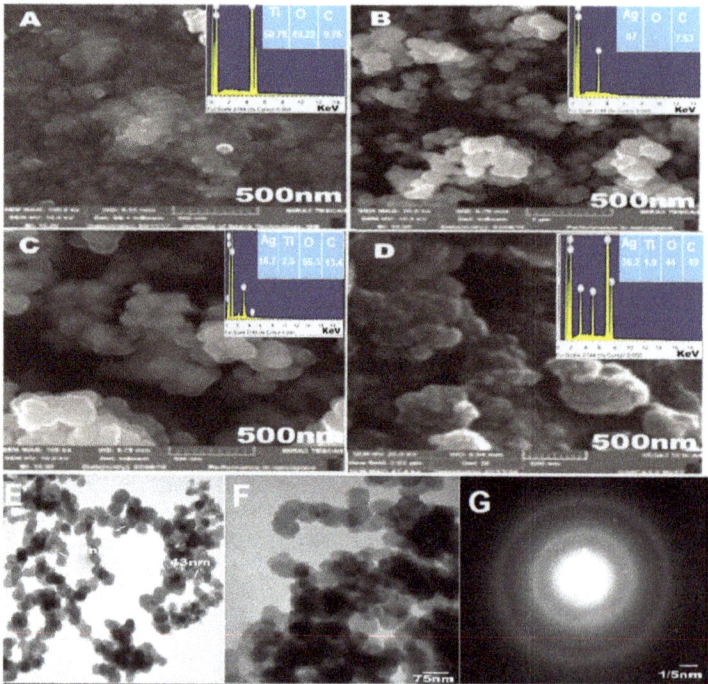

Figure 2. FESEM images of (**A**) TiO$_2$NPs, (**B**) AgNPs, (**C**) TiO$_2$/Ag nanocomposite, (**D**) CIP-TiO$_2$/Ag/CS nanohybrid TEM analysis, (**E**) TiO$_2$/Ag composites, (**F**) CIP-TiO$_2$/Ag/CS nanohybrids, and (**G**) selected area electron diffraction (SAED).

The TEM results confirmed the outcomes of FESEM analysis, as shown in Figure 2E,F, round and spherical morphology was depicted, while the purity of the newly developed CIP-TiO$_2$/Ag/CS nanohybrid was observed by SAED, as shown in Figure 2G. The particle sizes of the Ag/TiO$_2$ composite and CIP-TiO$_2$/Ag/CS nanohybrid were 47–75 nm and 20–80 nm, respectively [28]. The SAED pattern of the prepared CIP-Ag/TiO$_2$/CS nanohybrid demonstrated that Ag/TiO$_2$ contained a face-centered cubic crystalline phase. The SAED pattern further exhibited discrete circular diffraction rings corresponding to the anatase phase of TiO$_2$ NPs. Moreover, SAED showed less agglomeration, which supported the facts that phytocompounds of leaf extract of *M. concanensis* and chitosan contribute their role as previously explained in the literature by Senthilkumar et al., 2019 [41]. Obtained polycrystalline diffraction rings of fabricated spherical NPs were in agreement with previous studies reported by Senthilkumar et al., 2019, Hussein et al., 2021, and Mohamed et al., 2020 [41–43].

3.2.2. XRD, FTIR, and Zeta Potential Analysis of Synthesized Nanoformulations

The XRD pattern of TiO$_2$NPs (Figure 3(Aa)) represented the(101), (004), (200), (105), (211), and (204) plane indices that correspond to the crystalline anatase phase, as supported by (JCPDS No. 84-1285). The XRD peaks of AgNPs (Figure 3(Ab)) show the (111), (200), (220), and (311) crystallographic planes at 2θ° = 38.18°, 44.25°, 64.72°, and 77.40° leading to face-centered cubic metallic silver crystals [44]. It can be inferred that Ag ions (Ag$^+$) are strongly reduced by the *M. concanensis* leaf extract during the synthesis process. Any diffraction peak related to silver oxides was not observed. In the Ag/TiO$_2$ nanocomposite (c in the Figure 3A),the characteristic XRD peaks show the anatase phase of TiO$_2$ and the face-centered cubic silver content without any sign of any other diffraction peaks as impurity [28]. In CSNPs (d in the Figure 3A), the characteristic XRD peak was indicated at 21.8° crystallinity and the purity of chitosan in the nanostructureis in accordance with Zafar et al., 2020). The XRD profile of the Ag/TiO$_2$/CS nanohybrid (e in the Figure 3A) and the XRD profile of the CIP-Ag/TiO$_2$/CS nanohybrid (f in the Figure 3A) revealed the diffraction peaks of anatase phase of TiO$_2$, Ag and CS NPs. These characteristic peaks were in accordance with previously described peaks of Ag NPs by Lei et al., 2012 [44] while TiO$_2$ and CS peaks correspond to the findings of Zafar et al., 2020 [29]. It is observed that the diffraction peaks of this nanocarrier are shifted to a high angle region, which indicated that foreign material, i.e., CS and CIP, inserted the stress on the lattice of the host material (Ag/TiO$_2$). It is also noted that the leading peak of Ag at 38.18° overlapped with the peak of TiO$_2$ at 38° and suppressed the signal of TiO$_2$.

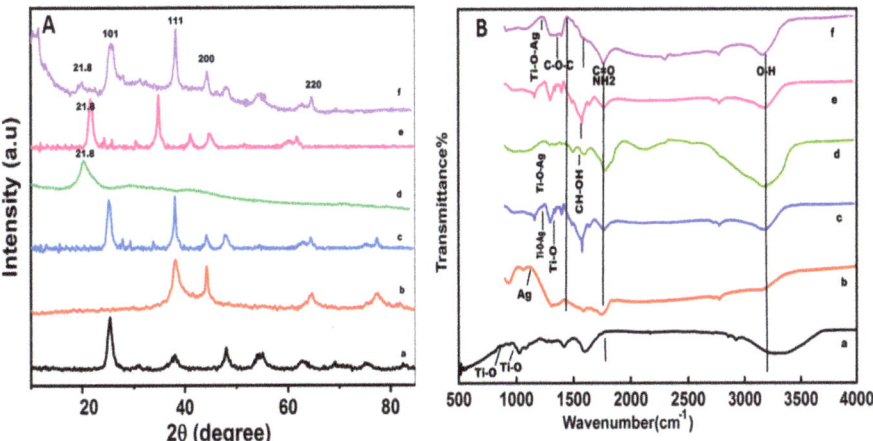

Figure 3. XRD profiles (**A**) and Fourier Transform Infrared spectroscopy (FTIR) spectrums of eco-friendly prepared nanoformulations (**B**) are shown, Key: (**a**) TiO$_2$ NPs, (**b**) AgNPs, (**c**) Ag/TiO$_2$ nanocomposite, (**d**) chitosan (CS) nanoparticles (NPs), (**e**) unloaded Ag/TiO$_2$/CS nanohybrid, and (**f**) CIP-Ag/TiO$_2$/CS nanohybrid.

The crystalline particle size of nanoformulationsis measured about the peaks centered at (101) of anatase TiO$_2$ [45] and (111) of Ag by using Scherrer's equation. The PS (particles size) and the crystalline size of the newly developed drug nanocarrier were 19 nm ± 1.98 and 0.9821 ± 0.76 Å respectively. The FTIR spectra of TiO$_2$NPs, AgNPs, CSNPs, Ag/TiO$_2$ nanocomposite, unloaded Ag/TiO$_2$/CS nanohybrids, and CIP-Ag/TiO$_2$/CS nanohybrids are displayed in Figure 3B. The FTIR spectrum of pure TiO$_2$ (Figure 3Ba) exhibited emerging characteristic peaks of absorption at 3408 cm^{-1} that belong to the superposition of the hydroxyl groups (O–H), which evidences the coordination of water molecule to Ti^{4+}cations. The absorption band cantered at 2928 cm^{-1} is assigned to C–H stretching vibrations. The signature at 1603 cm^{-1} can be attributed to C=O stretching vibrations due to the butyl group, organic species as starting precursor solutions, and adsorbed water molecules on the surface of the nanoformulations. The absorption band in the range of 766–610 cm^{-1} is related to the Ti-O bonding that authenticates the formation of TiO$_2$ [16]. The FTIR spectrum of AgNPs (b in the Figure 3B) revealed the characteristic peak at 3424 cm^{-1} corresponding to O−H stretching vibrations of adsorbed water molecules. The peaks at 2919 cm^{-1} and 2841 cm^{-1} indicated alkanes (C–C) stretching vibrations. The signature that appeared at 1625 cm^{-1} is attributed to the bending vibrations of the alkene group [46]. The peak at 1099 cm^{-1} was assigned to the asymmetric and symmetric C=O stretching vibrations due to the carbonyl group present in the leaf extraction. Alkanes, alkenes, and carbonyl groups of leaves extraction are mainly involved in the reduction of Ag$^+$ to AgNPs. The FTIR spectrum of the TiO$_2$/Ag nanocomposite (c in the Figure 3B) displayed band ranges in the region from 800 to 530 cm^{-1} that are attributed to Ti−O stretching mode and Ti−O−Ag/Ag−O−Tilinkage [40]. The FTIR spectrum of CS (Figure 3Bd) exhibited a high absorption peak of 3423 cm^{-1} and 1636 cm^{-1} due to the availability of a free –OH group from water molecules, an amino group, and a C=O carbonyl moiety group. The value at 1018 cm^{-1} corresponded to the throttle vibration of the C−O−C bond of epoxy or alkoxy. The signatures at 1269 cm^{-1} and 1419 cm^{-1} were due to C−O and CH−OH bonds [46]. The FTIR study of unloaded Ag/TiO$_2$/CS (e in the Figure 3B) characteristic peaks of metal components and CS differential peaks were prominently showed in the spectrum. Moreover, the binary junction of Ti–O and Ag metals was also displayed in the spectrum. The FTIR of CIP-Ag/TiO$_2$/CS nanohybrid (f in the Figure 2B) showed peaks at around 1010 cm^{-1} and 1600 cm^{-1} that are correlated with aromatic bending and stretching. It is clear that the absorption peak centered at 596 cm^{-1} is due to the metal oxygen metal (Ti−O−Ag) mode of vibration.

Zeta potential is the capacity of suspended particles to affect their stability, and the zeta potential greater than 30 mV or less than −30 mV can be distributed permanently in the medium [47]. The zeta potential values of AgNPs (Figure 4a, TiO$_2$NPs (Figure 4b), TiO$_2$/Ag nanocomposite (Figure 4c), and CSNPs (Figure 4d) were −110± 0.5 mV, −123 ± 1 mV, −200 ± 6 mV, and 35.12 ± 2.69 mV, respectively. The newly synthesized CIP-Ag/TiO$_2$/CS nanohybrid showed good stability with a zeta potential value of 67.45 ± 1.8 mV (Figure 4e).

It is worth mentioning that in relation to the adhesion of NPs with bacteria, the surface charge or zeta potential is crucial, since it was demonstrated that positively charged NPs will interact with negatively charged bacteria, and when these bacteria come near positively charged NPs, this ultimately led to the penetration and destruction of bacteria [48,49].

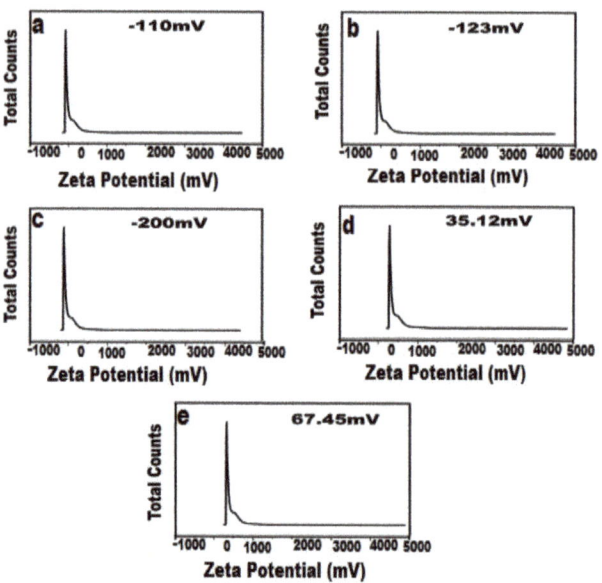

Figure 4. Zeta potential of biosynthesized nanoformulations, (**a**) TiO$_2$NPs, (**b**) AgNPs, (**c**) TiO$_2$/Ag nanocomposite, (**d**) CS, (**e**) CIP-TiO$_2$/Ag/CS nanohybrid.

3.3. Encapsulation Efficiency of CIP-TiO$_2$/Ag/CS Nanohybrid

The encapsulation efficiency of the CIP-TiO$_2$/Ag/CS nanohybrid in the CS system was found as 90% ± 2.07. Hanna and Saadalso reported a good encapsulation efficiency of CIP inside hydrogel made up of chitosan [50].

3.4. Antibacterial Activity of Greenly Synthesized Nanoformulations

Antibacterial activity was studied by the disc diffusion method, as shown in Figure 5. *E. coli* was considered to be MDR on the basis of a resistant pattern against synthetic antibiotics that are generally used for the cure of mastitis, as shown in Figure 5A. Ciprofloxacin was considered the most efficient drug, but according to the present study, it has lost its efficacy; *E. coli* strains have developed resistance against this drug and become super bugs by showing resistance against all the recommended values of MIC, as shown in Figure 5B by employing the E-strip method. The CIP-Ag/TiO$_2$/CS nanohybrid was found to be the most efficient and active antimicrobial agent; the means of zones of inhibition of *E. coli* produced by nanoformulations were measured. Table 1 shows the anti *E. coli* activity of synthesized nanoformulations. As shown in Table 1, the dose-dependent antibacterial property of chitosan, the Ag/TiO$_2$ nanocomposite, and the ciprofloxacin-loaded CS nanohybrid was observed. The CIP-Ag/TiO$_2$/CS nanohybrid exhibited the highest zone of inhibition of 23 mm ± 1.185 by using 0.2048 µg/mL of the CIP-Ag/TiO$_2$/CS nanohybrid, which is an admirable antibacterial activity.

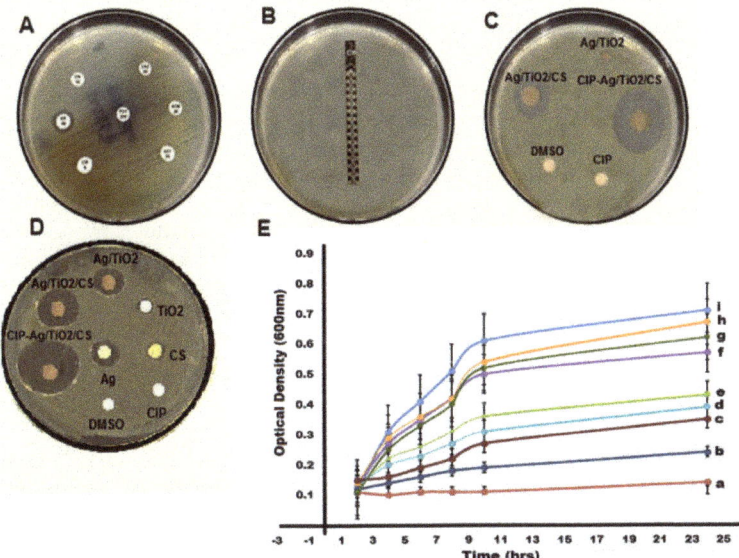

Figure 5. (**A**) Showing resistance pattern of *E. coli* as MDR pathogens, (**B**) displaying the highest minimum inhibitory concentration (MIC) value of ciprofloxacin with no zone by E-strip against *E. coli*, (**C**) zone of inhibition exhibited by synthetic and prepared antimicrobial agents, (**D**) zones of inhibition (ZIs) shown by synthesized nanoformulations at respective MICs, (**E**) kinetics of growth curves for MDR *E. coli* at 0.0512 µg/mL (MIC of nanohybrid), (**a**) NC (Negative control, pure NB), (**b**) CIP-Ag/TiO$_2$/CS nanohybrid, (**c**) unloaded Ag/TiO$_2$/CS, (**d**) Ag/TiO$_2$, (**e**) AgNPs, (**f**) TiO$_2$NPs, (**g**) CS, (**h**) CIP, (**i**) PC (positive control *E. coli* in broth).

Killing Kinetics of Nanoformulations against MDR *E. coli*

According to the current findings, the CIP-Ag/TiO$_2$/CS nanohybrid is highly effective in combating MDR *E. coli*, as reflected in Figure 5E, which presented a growth kinetic curve. The growth of MDR *E. coli* was ceased by the CIP-Ag/TiO$_2$/CS nanohybrid within 6–8 h of incubation by reducing the OD values close to the negative control of the study (autoclaved nutrient broth), as shown in Figure 5E. A study conducted by Li showed an antibacterial attack of Ag NPs, TiO$_2$ NPs, and Ag/TiO$_2$ nanocomposite against MDR *E. coli* [46], but the activity was not as remarkable as observed in this study by our synthesized nanoformulations; particularly, the CIP-Ag/TiO$_2$/CS nanohybrid halted the growth of *E. coli* within a few hours of exposure. The findings of our analysis showed a mutual antibacterial activity of ciprofloxacin with the Ag/TiO$_2$/CS composite. Shahverdi and colleagues observed improved antibiotic activity against the bacterial panel, using a combination of silver nanoparticles and FDA (Food and Drug Administration) proven antibiotics [51,52].

Table 1. Zone of inhibition and SD (Standard deviation) values of synthesized nanoformulations agents and synthetic antibiotic at various concentrations against MDR *E. coli*.

Antimicrobial Agents	Concentrations (µg/mL) and Zone of Inhibitions (mm)		
	MICs	MICsX2	MICsX3
Loaded CIP-TiO$_2$/Ag/CS	15 ± 1.06	18 ± 0.98	23 ± 1.185
Unloaded TiO$_2$/Ag/CS	7 ± 0.03	9 ± 0.10	10 ± 1.35
Ag/TiO$_2$ Nanocomposite	5 ± 0.12	7 ± 0.14	9 ± 1.76
TiO$_2$ NPs	2 ± 0.11	9 ± 1.05	11 ± 0.40
Ag NPs	3.5 ± 0.02	8 ± 1.13	12 ± 1.79
CS NPs	1 ± 0.17	3 ± 0.90	7 ± 0.64
CIP	0.9 ± 0.03	2 ± 0.48	5 ± 0.58
DMSO	-	-	-

3.5. MDR E. coli Cell Morphology Alterations Mediated by CIP-TiO$_2$/Ag/CS Nanohybrid

FESEM and TEM have been used to visualize the CIP-TiO$_2$/Ag/CS nanohybrid-induced potential morphological alterations on MDR *E. coli* using respective MIC concentration. After treating the bacterial cells with the CIP-Ag/TiO$_2$/CS nanohybrid, cytolysis in *E. coli* cells can be seen in Figures 6 and 7. In the control group, (untreated) cells shown intact, uniform, and plump morphology as seen in Figures 6A and 7A; however, after 6 h of treatment, the surface of the previously healthy *E. coli* cells showed deep rill-like folds, which led to the detachment of the membrane from the cell wall (Figures 6B and 7B). Almost all cells have low-density regions in their center, which clearly indicates that cytoplasm was damaged by the nanohybrid and the outer membrane was disintegrated, but the cytoplasmic shape was still maintained (Figures 6B and 7B). CIP alone was tested against *E. coli* strains and results were concluded in Figure 6D–F by FESEM and TEM (Figure 7D–F).

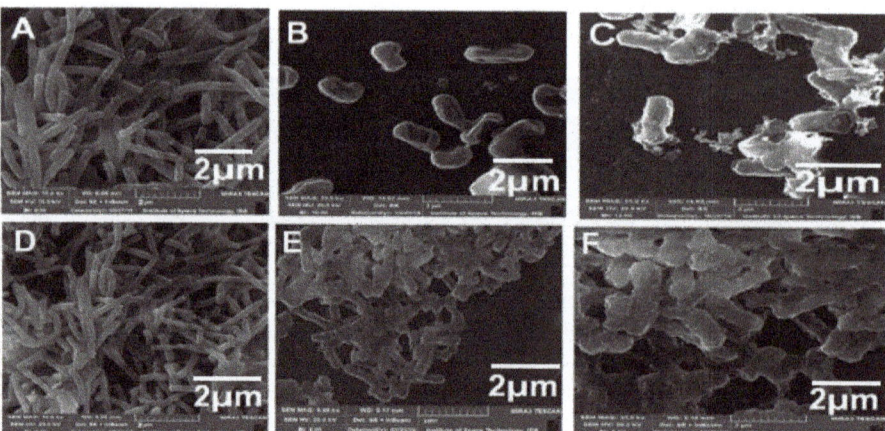

Figure 6. FESEM micrographs displaying morphological changes in MDR *E. coli* cells treated with CIP-Ag/TiO$_2$/CS nanohybrid at MIC (0.0512 µg/mL) and at different intervals of time. (**A**) Untreated MDR *E. coli* cells, (**B**) MDR *E. coli* cells after 6 h of incubation, (**C**) MDR *E. coli* cells after 12 h of incubation, (**D**) *E. coli* with CIP at 0 h of incubation, (**E**) CIP at 6 h, and (**F**) CIP at 12 h culture of *E. coli*.

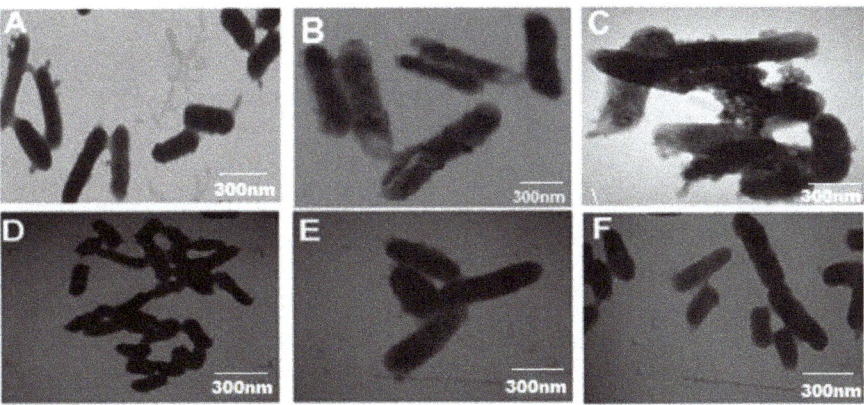

Figure 7. TEM micrographs displaying ultrastructural changes in MDR *E. coli* cells treated with CIP-Ag/TiO$_2$/CS nanohybrid at MIC (0.0512 µg/mL) and at different intervals of time. (**A**) Untreated MDR *E. coli* cells, (**B**) MDR *E. coli* cells after 6 h of incubation, (**C**) MDR *E. coli* cells after 12 h of incubation, (**D**) *E. coli* with CIP at 0 h of incubation (**E**) CIP at 6 h, and (**F**) CIP at 12 h of *E. coli* treatment.

3.6. Live/Dead Assessment of CIP-Ag/TiO$_2$/CS Nanohybrid-Treated Bacteria

Flow cytometry allows the rapid identification of drug-conjugated nanoparticles internalization in live cells [53]. *E. coli* exhibited increased cell membrane damage and cell inclusion leaking when treated with the CIP-Ag/TiO$_2$/CS nanohybrid, as confirmed by flow cytometry analysis. Annexin V-FITC and PI dyes were used to reveal stages of apoptosis in the *E. coli* cells after interacting with CIP-Ag/TiO$_2$/CS nanohybrid. Phosphatidylserin is a phospholipid abundant in the internal surface of the plasma membrane that is exposed to calcium-dependent signals in the outer leaflet during early apoptosis. PI, an intact impermeable dye can only pass via the cells until it is weakened or dead. In connection with PI, we have obtained a rapid and reliable analysis of cellular structural damage based on a flow cytometric analysis. Our result showed that the cell population in the right upper quadrant in Figure 8B, which indicates late apoptotic cells in comparison to Figure 8A without treatment in which 99.81% cells were live. Current finding semphasizing on the efficacy of nanohybrid by displaying 67.87% of late apoptotic cells and 32.13% cells showed early apoptosis (Figure 8B). Late apoptosis occurred when the nanohybrid penetrated in the *E. coli* cells instigated death, as PI dye is permeable to dead cells only [54]. The *E. coli* cells in the late apoptotic stage were sensitive to Annexin V-FITC dye showing the damaged cell membrane, and the leakage of content was confirmed by PI binding. We have confirmed that cell death caused by CIP-Ag/TiO$_2$/CS nanohybrid displays late apoptotic attributes. Cohesively, data show that the CIP-Ag/TiO$_2$/CS nanohybrid increased the permeation of the *E. coli* cells, potentially leading to cell death.

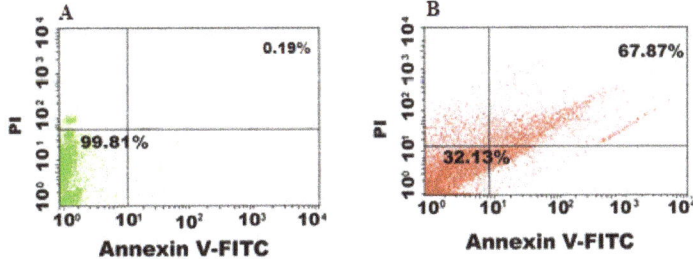

Figure 8. Flow cytometer data nanohybrid-induced cell death. (**A**) control (untreated), (**B**) CIP-TiO$_2$/Ag/CS nanohybrid-treated *E. coli* cells.

3.7. Ex Vivo Drug Release Study CIP-Ag/TiO₂/CS Nanohybrid

The exvivo drug release profile of CIP from the CIP-Ag/TiO$_2$/CS nanohybridis presented in Figure 9. The CIP-Ag/TiO$_2$/CS nanohybrid demonstrated the cumulative drug release (89 ± 2.43) at 8hrs of incubation, which was compared to the control value of CIP (94 ± 1.97) during 24 h. Previous study correlates with our findings; slow and sustained release of drug from the cross-linked structure of polymers enclosing antibiotics was reported [55]. The important requirements for designing an efficient delivery system areto ensure the continued and sustained release of the encapsulated drug to the biological system [56]. Our result has confirmed that the drug conjugated with Ag/TiO$_2$/CS nanoparticles and encapsulated in chitosan nanoparticles has improved stability in the acid medium and sustained and prolonged release of the drug. This stability and regulated release is due to chitosan encapsulation and most possibly due to the hydrophilic nature of chitosan. The antibiotic ciprofloxacin was trapped within the polymer matrixof chitosan, which supported the slow release of drugs through the diffusion process. Taken together, our results showed a burst CIF release at 8 h, followed by a sustained release in next 24 h. Shah and co-workers have documented the continued release of moxifloxacin from a nanocomposite of chitosan, and their findings are in agreement with our results [57].

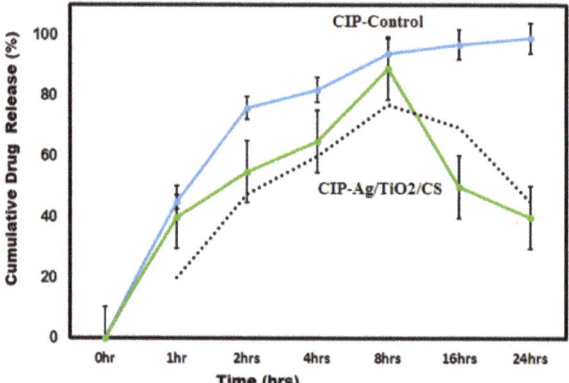

Figure 9. Ex vivo drug release profile of CIP-Ag/TiO$_2$/CS nanohybrid in phosphate buffer solution (PBS, pH 7.4, 37 °C).

3.8. Ex Vivo Cytotoxicity of CIP-Ag/TiO₂/CS Nanohybrid on Mammalian Cell Lines and Human RBCs

Since most nanoparticles are semi-synthetic or totally synthetic, in vivo toxicity is critical. Particularly if the nanoparticle contains an antibiotic in a formulation that can affect the normal metabolism of antibiotics, this may reduce the toxicity profile of the antibiotic and allow safer use of these drugs. The potential cytotoxicity at various concentrations (0.02, 0.1, and 0.2 μg/mL) was being analyzed on primary cultures of proliferating bovine mammary gland epithelial cells (BMGE). BMGE cells were found metabolically viable and proliferating after treatment, in a dose-dependent manner with synthesized nanoformulations, which were Ag NPs, TiO$_2$ NPs, TiO$_2$/Ag nanocomposites, CS NPs, pure CIP, Ag/TiO$_2$/CS nanohybrids, and CIP-Ag/TiO$_2$/CS nanohybrids. The cells treated with PBS were used as a negative control, and celecoxib drug was used as positive control. As shown in Figure 10, the viability of BMGE cells was calculated to be 95.23% at lowest concentration (0.02 μg/mL) while 93.08% of cells were viable at the highest concentration of the CIP-Ag/TiO$_2$/CS nanohybrid (0.2 μg/mL). The viability pattern using different concentrations of synthesized nanoformulations are exhibited in Figure 10. Similar results were reported, where chitosan was considered as a safe therapeutic agent due to its nontoxic effect on cell lines [58]. The current study deep-rooted the biocompatibility of the synthesized nanoformulation for BMGE cells by displaying a good viability pattern. Tomankova reported the cytotoxic effect of Ag NPs in comparison with TiO_2 NPs [59], but in our study, chitosan nanoparticles encapsulation masks the cytotoxic effects of these metals.

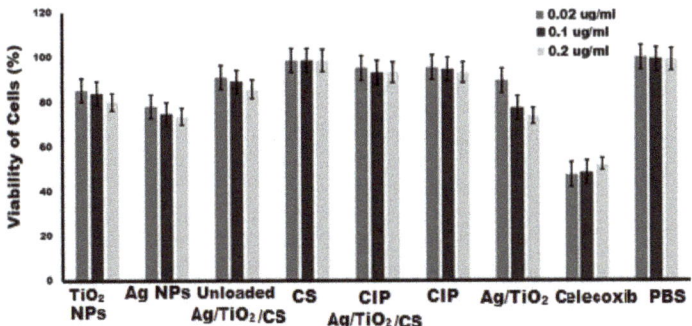

Figure 10. Cytotoxicity analysis of nanoformulations at various concentrations for 24 h of incubation, Celecoxib (PC) and PBS (NC) of study.

The good viability of BMGE cells in the presence of the CIP-Ag/TiO_2/CS nanohybrid is likely due to the indirect exposure of cells to Ag/TiO_2, which is safeguarded by a biocompatible polymer of chitosan nanoparticles, whereas ciprofloxacin is an already approved safe drug.

Blood tissue encounters directly or indirectly with nanoparticles and is able to transport nanoparticles to other cells, tissues, and organs. For this reason, it is highly required to study the toxicity on blood, mainly erythrocytes. In this study, hemolysis assay was designed to study the toxicity on RBCs after exposure to all green synthesized nanoformulations at various concentrations (0.02, 0.1, and 0.2 μg/mL). Nanoformulations allowed interacting for 6 h, and the findings are displayed in Figure 11. Our results revealed that RBCs remained viable after exposure to Ag NPs, TiO_2, Ag/TiO_2, unloaded Ag/TiO_2/CS NPs, pure CIP, and CS NPs; 98.2% viability was observed in CIP-Ag/TiO_2/CS nanohybrid treated red blood cells. The hemotoxicity of TiO_2 NPs was also studied by Li et al., 2008 [60], and the concentration and size-dependent toxicity of Ag NPs was evaluated by Choi et al., 2011 [61]. Hence, newly synthesized CIP-Ag/TiO_2/CS nanohybrids showed negligible hemolysis of RBCS and the proposed hemocompatibility, biocompatibility, and nontoxicity of the CIP-Ag/TiO_2/CS nanohybrid for both human and animal cells.

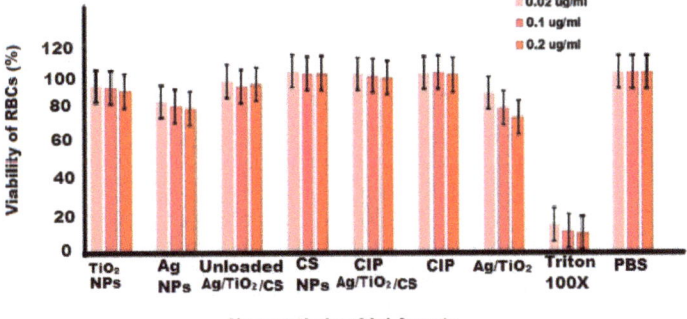

Figure 11. Hemolysis study on human red blood cells (RBCs) with nanoformulations at various concentrations, Triton 100X (PC) and PBS (NC) of study.

3.9. CIP-TiO$_2$/Ag/CS Nanohybrid-Mediated Antibacterial Activity Mechanism

Taken together, our data underline a possible mechanism mediated by the CIP-Ag/TiO$_2$/CS nanohybrid that involved an initial step of adhesion of the particles to the cell wall, which was followed by its destruction of the cell wall (detachment from the outer membrane) to penetrate and disrupt the cell integrity, ultimately leading to apoptotic cell death. Thereby, compared to CIP alone, the enhanced bactericidal effect could be attributed to a synergistic effect of CIP with the TiO$_2$/Ag nanocomposite; moreover, it also showed the surface charges modification of CIP-Ag/TiO$_2$/CS nanohybrid by the biocompatible CS doping. Previous studies [15,62] reported that TiO$_2$/Ba hybrid nanoclusters effectively reduced the cell count for both Gram-positive and Gram-negative bacteria. It was reported that CS can enhance the permeability of the cell membrane via the interaction of anionic groups on the cell [56]. The presence of the –NH bond is important for antimicrobial activity of CS [63]. Then, CIP anchored to CS triggers the liberation of Ag$^+$ from the nanohybrid material, enhancing Ag$^+$ penetration/entrance into the cell through the cell membrane, and subsequently enhancing the ROS levels, leading to cell death [64]. Meanwhile, it is known that TiO$_2$ further disrupts the barrier properties of the outer membrane of the bacteria by ROS [60,64]. CIP is an exceeding energetic antibiotic against diverse microorganisms and effectively causes double-stranded DNA (dsDNA) breaks and inhibits the DNA gyrase [65]. Eventually, the synergistic action of each entity of the newly biosynthetized nanohybrid has led to the disruption of the cell membrane and ROS-mediated oxidative stress, resulting in enhanced antibacterial activity (Figure 12).

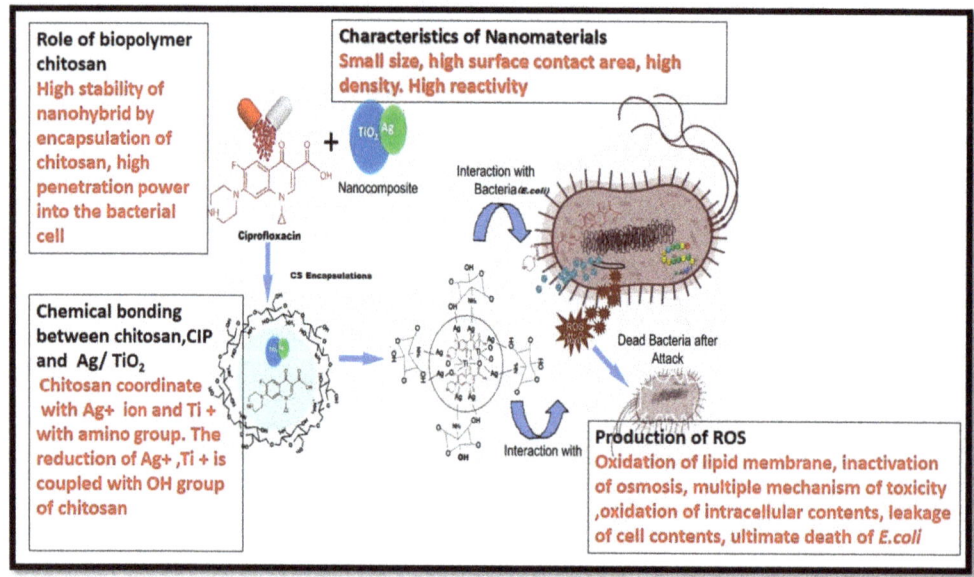

Figure 12. Putative CIP-TiO$_2$/Ag/CS nanohybrid-mediated cellular and molecular bactericidal mechanism.

The newly developed CIP-Ag/TiO$_2$/CS nanohybrid has been prepared by using a unique leaf extract of *Moringa concanensis*, which has been less explored for the synthesis of nanoparticles. This newly developed CIP-Ag/TiO$_2$/CS nanohybrid is not reported to treat mastitis caused by MDR *E. coli*. The nanoparticles obtained in the study had a small particle size (20–40 nm) and suitable polarity (67.45 ± 1.8 mV) due to chitosan encapsulation, which may increase the drug penetration into the MDR *E. coli* cells, as evident by flow cytometry and improve its antibacterial activity to overcome the resistance mechanism. The results showed that the CIP-Ag/TiO$_2$/CS nanohybrid can successfully inhibit the growth of MDR *E. coli* strikingly with an MIC value lower than the MIC of ciprofloxacin itself. It was anticipated that the CIP-Ag/TiO$_2$/CS nanohybrid could be applied broadly in the treatment of livestock infectious diseases (mastitis) as an alternative therapeutic agent in the field of medicine due to highly biocompatibilities.

4. Conclusions and Perspectives

This research was conducted to improve the effectiveness of existing groups of antibiotics (ciprofloxacin) and to reduce dosage and mitigate the associated toxicity. This study illustrates the successful development of a CIP-Ag/TiO$_2$/CS nanohybrid. The eco-friendly synthesized CIP-Ag/TiO$_2$/CS nanohybrid exerted excellent antibacterial activity at relatively very low MIC compared to CIP alone. The synthesis of Ag/TiO$_2$ occurred through reduction of *M. concanensis* leaves extract followed by an ionic gelation method for the conjugation of CIP and CS encapsulation. The excellent rapid anti *E. coli* activity was exhibited by our synthesized nanohybrid formulation. Flow cytometry revealed cell membrane damage leading to cell lysis as confirmed by SEM and TEM, as major morphological alterations were seen in *E. coli* cells. The encapsulation efficiency of the CIP-TiO$_2$/Ag/CS nanohybrid in the CS matrix was calculated 90% ± 2.07. Drug released kinetics exhibited sustained drug release from the nanohybrid. The nanohybrid was proved to be safe and nontoxic on bovine mammary gland epithelial cells. Further in vivo experiments should be performed to investigate the biocompatibility and undesired adverse effects for their possible therapeutic use. The use of available antibiotics conjugated or attached to nanoparticles offers an alternate angle to antibiotic therapy.

Author Contributions: Conceptualization, B.U.; and N.Z.; methodology, M.B.K.N.; software, N.J.; validation, G.S., and W.-U.-N.; formal analysis, F.M., N.J., A.B.; investigation, N.Z. and F.M.; resources, N.J., S.S.; data interpretation, N.Z. and F.M.; writing—original draft preparation, N.Z.; review and editing, B.U. and F.M.; visualization, M.B.K.N.; supervision, B.U.; project administration, M.B.K.N. All authors have read and agreed to the published version of the manuscript.

Funding: This research received no external funding.

Institutional Review Board Statement: The present study was approved by Institution Review Board and Ethical Review Board of International Islamic University, Islamabad, Pakistan (Reg #22-FBAS/PHDBT/F-14).

Informed Consent Statement: Consent was obtained from all volunteers involved in the study, informed briefly that the provided blood samples will be exclusively processed for research purpose only.

Data Availability Statement: Article contains all the related data and information.

Conflicts of Interest: The authors declare no conflict of interest.

References

1. Kimera, Z.I.; Mshana, S.E.; Rweyemamu, M.M.; Mboera, L.E.G.; Matee, M.I.N. Antimicrobial use and resistance in food-producing animals and the environment: An African perspective. *Antimicrob. Resist. Infect. Control* **2020**, *37*, 1–12. [CrossRef]
2. Ventola, C.L. The Antibiotic Resistance Crisis: Part 1: Causes and Threats. *Pharm. Ther.* **2015**, *40*, 277–283.
3. Pereira, R.V.V.; Siler, J.D.; Bicalho, R.C.; Warnick, L.D. In vivo selection of resistant *E. coli* after ingestion of milk with added drug residues. *PLoS ONE* **2014**, *9*, e115223. [CrossRef]
4. Abid, S.; Uzair, B.; Niazi, M.B.H.; Fasim, F.; Bano, S.A.; Jamil, N.; Batool, R.; Sajjad, S. Bursting the virulence traits of MDR Strains of *Candida albicans* using sodium alginate based microspheres containing nystatin loaded MgO/CuO nanocomposites. *Int. J. Nanomed.* **2021**, *16*, 1157–1174. [CrossRef]
5. Pereira, V.V.R.; Lima, S.; Siler, J.D.; Foditsch, C.; Warnick, L.D.; Bicalho, R.C. Ingestion of milk containing very low concentration of antimicrobials: Longitudinal effect on fecal microbiota composition in preweaned calves. *PLoS ONE* **2016**, *11*, e0147525.
6. Agga, G.E.; Arthur, T.M.; Durso, L.M.; Harhay, D.M.; Schmidt, J.W. Antimicrobial-resistant bacterial populations and antimicrobial resistance genes obtained from environments impacted by livestock and municipal waste. *PLoS ONE* **2015**, *10*, e0132586. [CrossRef]

7. World Health Organization. Global Priority List of Antibiotic-Resistant Bacteria to Guide Research, Discovery, and Development of New Antibiotics. 2017. Available online: http://www.who.int/medicines/publications/global-priority-list-antibiotic-resistantbacteria/en/ (accessed on 27 February 2017).
8. Anderson, V.E.; Gootz, T.D.; Osheroff, N. Topoisomerase IV catalysis and the mechanism of quinolone action. *J. Biol. Chem.* **1998**, *273*, 17879–17885. [CrossRef] [PubMed]
9. Van der Putten, B.C.L.; Remondini, D.; Pasquini, G.; Janes, V.A.; Matamoros, S.; Schultsz, C. Quantifying the contribution of four resistance mechanisms to ciprofloxacin MIC in *Escherichia coli*: A systematic review. *J. Antimicrob. Chemother.* **2018**, *74*, 298–310. [CrossRef]
10. Kumar, M.; Curtis, A.; Hoskins, C. Application of Nanoparticle Technologies in the Combat against Anti-Microbial Resistance. *Pharmaceutics* **2018**, *10*, 11. [CrossRef] [PubMed]
11. Slavin, Y.N.; Asnis, J.; Häfeli, U.O.; Bach, H. Metal nanoparticles: Understanding the mechanisms behind antibacterial activity. *J. Nanobiotechnol.* **2017**, *15*, 1–20. [CrossRef] [PubMed]
12. Kandi, V.; Kandi, S. Antimicrobial properties of nanomolecules: Potential candidates as antibiotics in the era of multi-drug resistance. *Epidemiol. Health* **2015**, *37*, e2015020. [CrossRef]
13. Gnanaprakasam, A.; Sivakumar, V.; Sivayogavalli, P.; Thirumarimurugan, M. Characterization of TiO_2 and ZnO nanoparticles and their applications in photocatalytic degradation of azodyes. *Ecotox. Environ. Saf.* **2015**, *121*, 121–125. [CrossRef]
14. Kravanja, G.; Primožič, M.; Knez, Ž.; Leitgeb, M. Chitosan-Based (Nano) Materials for Novel Biomedical Applications. *Molecules* **2019**, *24*, 1960. [CrossRef]
15. Marta, B.; Potara, M.; Iliut, M.; Jakab, E.; Radu, T.; Imre-Lucaci, F.; Katona, G.; Popescu, O.; Astilean, S. Designing chitosan–silver nanoparticles–graphene oxide nanohybrids with enhanced antibacterial activity against *Staphylococcus aureus*. *Colloids Surf. A Physicochem. Eng. Asp.* **2015**, *487*, 113–120. [CrossRef]
16. Elgadir, M.A.; Uddin, M.S.; Ferdosh, S.; Adam, A.; Chowdhury, A.J.K.; Sarker, M.Z.I. Impact of chitosan composites and chitosan nanoparticle composites on various drug delivery systems: A review. *J. Food Drug Anal.* **2015**, *23*, 619–629. [CrossRef]
17. Thu, H.P.; Nam, N.H.; Quang, B.T.; Son, H.A.; Toan, N.L.; Quang, D.T. In vitro and in vivo targeting effect of folate decorated paclitaxel loaded PLA–TPGS nanoparticles. *Saudi Pharm. J.* **2015**, *23*, 683–688. [CrossRef]
18. Zhang, X.F.; Liu, Z.G.; Shen, W.; Gurunathan, S. Silver nanoparticles: Synthesis, characterization, properties, applications, and therapeutic approaches. *Int. J. Mol. Sci.* **2016**, *17*, 1534. [CrossRef]
19. Basiuk, V.A.; Basiuk, E.V. *Green Processes for Nanotechnology; From Inorganic to Bioinspiring Nanomaterials*; Springer International Publishing: Cham, Switzerland, 2015; pp. 35–73.
20. Anbazhakan, S.; Dhandapani, R.; Anandhakumar, P.; Balu, S. Traditional Medicinal Knowledge on *Moringa concanensis* Nimmo of Perambalur District, Tamilnadu. *Anc. Sci. Life* **2007**, *26*, 42–45. [PubMed]
21. Balamurugan, V.; Balakrishnan, V. Evaluation of phytochemical, pharmacognostical and antimicrobial activity from the bark of *Moringa concanensis* Nimmo. *Int. J. Curr. Microbiol. Appl. Sci.* **2013**, *2*, 117–125.
22. Unnerstad, H.E.; Lindberg, A.; Waller, K.P.; Ekman, T.; Artursson, K.; Nilsson-Öst, M.; Bengtsson, B. Microbial aetiology of acute clinical mastitis and agent-specific risk factors. *Vet. Microbiol.* **2009**, *137*, 90–97. [CrossRef]
23. Morrissey, I.; Bouchillon, S.K.; Hackel, M.; Biedenbach, D.J.; Hawser, S.; Hoban, D.; Badal, R.E. Evaluation of the Clinical and Laboratory Standards Institute phenotypic confirmatory test to detect the presence of extended-spectrum β-lactamases from 4005 *Escherichia coli*, *Klebsiellaoxytoca*, *Klebsiella pneumoniae* and *Proteus mirabilis* isolates. *J. Med. Microbiol.* **2014**, *63*, 556–561. [CrossRef] [PubMed]
24. Barker, K. *At the Bench: A Laboratory Navigator*; Cold Spring Harbor Laboratory Press: Cold Spring Harbor, NY, USA, 1998.
25. Yadav, M.; Kaur, P. A review on exploring phytosynthesis of silver and gold nanoparticles using genus Brassica. *Int. J. Nanopart.* **2018**, *10*, 165–177. [CrossRef]
26. Shameli, K.; Ahmad, M.B.; Zargar, M.; Yunus, W.M.; Rustaiyan, A.; Ibrahim, N.A. Synthesis of silver nanoparticles in montmorillonite and their antibacterial behavior. *Int. J. Nanomed.* **2011**, *6*, 581–590. [CrossRef] [PubMed]
27. Jamil, B.; Habib, H.; Abbasi, S.A.; Ihsan, A.; Nasir, H.; Imran, M. Development of cefotaxime impregnated chitosan as nano-antibiotics: De novo strategy to combat biofilm forming multi-drug resistant pathogens. *Front. Microbiol.* **2016**, *7*, 330. [CrossRef] [PubMed]
28. Yang, X.H.; Fu, H.T.; Wang, X.C.; Yang, J.L.; Jiang, X.C.; Yu, A.B. Synthesis of silver-titanium dioxide nanocomposites for antimicrobial applications. *J. Nanopart. Res.* **2014**, *16*, 2526. [CrossRef]
29. Zafar, N.; Uzair, B.; Niazi, M.B.K.; Sajjad, S.; Samin, G.; Arshed, M.J.; Rafiq, S. Fabrication & characterization of chitosan coated biologically synthesized TiO_2 nanoparticles against PDR *E. coli* of veterinary origin. *Adv. Polym. Technol.* **2020**, *2020*, 8456024.
30. Liu, P.; Dai, Y.-N.; Zhang, J.-P.; Wang, A.-Q.; Wei, Q. Chitosan-alginate nanoparticles as a novel drug delivery system for nifedipine. *Int. J. Biomed. Sci.* **2008**, *4*, 221.
31. Jorgensen, J.H.; Turnidge, J.D. Susceptibility test methods: Dilution and disk diffusion methods. In *Manual of Clinical Microbiology*, 11th ed.; Jorgensen, J., Pfaller, M., Carroll, K., Funke, G., Landry, M., Richter, S., Warnock, D., Eds.; ASM Press: Washington, DC, USA, 2015; Chapter 71; pp. 1253–1273.
32. EUCAST. European Committee on Antimicrobial Susceptibility Testing. Breakpoint Tables for Interpretation of MICs and Zone Diameters Version 6.0. 2016. Available online: www.eucast.org (accessed on 10 January 2017).

33. Uzair, B.; Ahmed, N.; Ahmad, V.U.; Mohammad, F.V.; Edwards, D.H. The isolation, puri¢cation and biological activity of a novel antibacterial compound produced by *Pseudomonas stutzeri*. *FEMS Microbiol. Lett.* **2008**, *279*, 243–250. [CrossRef]
34. Berney, M.; Hammes, F.; Bosshard, F.; Weilenmann, H.U.; Egli, T. Assessment and Interpretation of Bacterial Viability by Using the LIVE/DEAD BacLight Kit in Combination with Flow Cytometry. *Appl. Environ. Microbiol.* **2007**, *73*, 3283–3290. [CrossRef]
35. Khan, B.A.; Ullah, S.; Khan, M.K.; Alshahrani, S.M.; Braga, V.A. Formulation and evaluation of Ocimumbasilicum-based emulgel for wound healing using animal model. *Saudi Pharm. J.* **2020**, *28*, 1842–1850. [CrossRef]
36. Caputo, F.; Mameli, M.; Sienkiewicz, A.; Licoccia, S.; Stellacci, F.; Ghibelli, L.; Traversa, E. A novel synthetic approach of cerium oxide nanoparticles with improved biomedical activity. *Sci. Rep.* **2017**, *7*, 1–13. [CrossRef] [PubMed]
37. Dobrovolskaia, M.A.; Clogston, J.D.; Neun, B.W.; Hall, J.B.; Patri, A.K.; McNeil, S.E. Method for analysis of nanoparticle hemolytic properties in vitro. *Nano Lett.* **2008**, *8*, 2180–2187. [CrossRef] [PubMed]
38. Dudley, M.N.; Ambrose, P.G.; Bhavnani, S.M.; Craig, W.A.; Ferraro, M.J.; Jones, R.N.; Clinical, Antimicrobial Susceptibility Testing Subcommittee of the Clinical and Laboratory Standards Institute. Background and rationale for revised Clinical and Laboratory Standards Institute interpretive criteria (breakpoints) for Enterobacteriaceae and *Pseudomonas aeruginosa*: I. Cephalosporins and aztreonam. *Clin. Infect. Dis.* **2013**, *56*, 1301–1309. [PubMed]
39. CLSI. *M100-S26: Performances Standards for Antimicrobial Susceptibility Testing*; Twenty-Fourth Informational Supplement; Clinical Laboratory Standards Institute: Wayne, PA, USA, 2016.
40. Bokare, A.; Sanap, A.; Pai, M.; Sabharwal, S.; Athawale, A.A. Antibacterial activities of Nd doped and Ag coated TiO_2 nanoparticles under solar light irradiation. *Colloids Surf. B Biointerfaces* **2013**, *102*, 273–280. [CrossRef]
41. Senthilkumar, P.; Yaswant, G.; Kavitha, S.; Chandramohan, E.; Kowsalya, G.; Vijay, R.; Sudhagar, B.; Kumar, D.R.S. Preparation and characterization of hybrid chitosan-silver nanoparticles (Chi-Ag NPs); A potential antibacterial agent. *Int.J. Biol. Macromol.* **2019**, *141*, 290–298. [CrossRef]
42. Hussein, E.M.; Desoky, W.M.; Hanafy, M.F.; Guirguis, O.W. Effect of TiO_2 nanoparticles on the structural configurations and thermal, mechanical, and optical properties of chitosan/TiO_2 nanoparticle composites. *J. Phys. Chem. Solids* **2021**, *152*, 109983. [CrossRef]
43. Mohamed, N. Synthesis of Hybrid Chitosan Silver Nanoparticles Loaded with Doxorubicin with Promising Anti-cancer Activity. *Biol. Nano Sci.* **2020**, *10*, 758–765. [CrossRef]
44. Georgekutty, R.; Seery, M.K.; Pillai, S.C. A highly efficient Ag-ZnOphotocatalyst: Synthesis, properties, and mechanism. *J. Phys. Chem. C* **2008**, *112*, 13563–13570. [CrossRef]
45. Lee, J.H.; Jung, K.Y.; Park, S.B. Modification of titania particles by ultrasonic spray pyrolysis of colloid. *J. Mater. Sci.* **1999**, *34*, 4089–4093. [CrossRef]
46. Li, J.; Xie, B.; Xia, K.; Li, Y.; Han, J.; Zhao, C. Enhanced antibacterial activity of silver doped titanium dioxide-chitosan composites under visible light. *Materials* **2018**, *11*, 1403. [CrossRef] [PubMed]
47. Wiącek, A.E.; Gozdecka, A.; Jurak, M. Physicochemical characteristics of chitosan–TiO_2 biomaterial. 1. Stability and swelling properties. *Ind. Eng. Chem. Res.* **2018**, *57*, 1859–1870. [CrossRef]
48. Hoseinzadeh, E.; Makhdoumi, P.; Taha, P.; Hossini, H.; Stelling, J.; Amjad Kamal, M. A review on nano-antimicrobials: Metal nanoparticles, methods and mechanisms. *Curr. Drug Metab.* **2017**, *18*, 120–128. [CrossRef] [PubMed]
49. Wang, L.; Hu, C.; Shao, L. The antimicrobial activity of nanoparticles: Present situation and prospects for the future. *Int. J. Nanomed.* **2017**, *12*, 1227. [CrossRef] [PubMed]
50. Hanna, D.H.; Saad, G.R. Encapsulation of ciprofloxacin within modified xanthan gum- chitosan based hydrogel for drug delivery. *Bioorg. Chem.* **2019**, *84*, 115–124. [CrossRef] [PubMed]
51. Shahverdi, A.; Fakhimi, A.; Shahverdi, H.; Minaian, S. Synthesis and effect of silver nanoparticles on the anti-bacterial activity of different antibiotics against *Staphylococcus aureus* and *Escherichia coli*. *Nanomed. Nanotechnol. Biol. Med.* **2007**, *3*, 168–171. [CrossRef]
52. Ashmore, D.A.; Chaudhari, A.; Barlow, B.; Barlow, B.; Harper, T.; Vig, K.; Miller, M.; Singh, S.; Nelson, E.; Pillai, S. Evaluation of *E. coli* inhibition by plain and polymer-coated silver nanoparticles. *Rev. Inst. Med. Trop. São Paulo* **2018**, *60*, e18. [CrossRef] [PubMed]
53. Carter, E.A.; Frank, E.P.; Hunter, P.A. Cytometric evaluation of antifungal agents. In *Flow Cytometry in Microbiology*; Lloyd, D., Ed.; Springer: London, UK, 1993; pp. 111–120.
54. Vanhauteghem, D.; Audenaert, K.; Demeyere, K.; Hoogendoorn, F.; Janssens, G.P.; Meyer, E. Flow cytometry, a powerful novel tool to rapidly assess bacterial viability in metal working fluids: Proof-of-principle. *PLoS ONE* **2019**, *14*, e0211583. [CrossRef]
55. Cui, Z.; Zheng, Z.; Lin, L.; Si, J.; Wang, Q.; Peng, X.; Chen, W. Electrospinning and crosslinking of polyvinyl alcohol/chitosan composite nanofiber for transdermal drug delivery. *Adv. Polym. Technol.* **2018**, *37*, 1917–1928. [CrossRef]
56. Shahriar, S.; Mondal, J.; Hasan, M.N.; Revuri, V.; Lee, D.Y.; Lee, Y.-K. Electrospinningnanofibers for therapeutics delivery. *Nanomaterials* **2019**, *9*, 532. [CrossRef]
57. Shah, A.; Buabeid, M.A.; Arafa, E.-S.A.; Hussain, I.; Li, L.; Murtaza, G. The wound healing and antibacterial potential of triple-component nanocomposite (chitosan-silver-sericin) films loaded with moxifloxacin. *Int. J. Pharm.* **2019**, *564*, 22–38. [CrossRef]
58. Campos, D.A.M.; Diebold, Y.; Carvalho, E.L.; Sánchez, A.; Alonso, M.J. Chitosan nanoparticles as new ocular drug delivery systems: In vitro stability, in vivo fate, and cellular toxicity. *Pharm. Res.* **2004**, *21*, 803–810. [CrossRef]

59. Tomankova, K.; Horakova, J.; Harvanova, M.; Malina, L.; Soukupova, J.; Hradilova, S.; Kejlova, K.; Malohlava, J.; Licman, L.; Dvorakova, M.; et al. Cytotoxicity, cell uptake and microscopic analysis of titanium dioxide and silver nanoparticles in vitro. *Food Chem. Toxicol.* **2015**, *82*, 106–115. [CrossRef]
60. Li, S.-Q.; Zhu, R.-R.; Zhu, H.; Xue, M.; Sun, X.-Y.; Yao, S.-D.; Wang, S.-L. Nanotoxicity of TiO_2 nanoparticles to erythrocyte in vitro. *Food Chem. Toxicol.* **2008**, *46*, 3626–3631. [CrossRef] [PubMed]
61. Choi, J.; Reipa, V.; Hitchins, V.M.; Goering, P.L.; Malinauskas, R.A. Physicochemical characterization and invitro hemolysis evaluation of silver nanoparticles. *Toxicol. Sci.* **2011**, *123*, 133–143. [CrossRef] [PubMed]
62. Vijayalakshmi, K.; Sivaraj, D. Synergistic antibacterial activity of barium doped TiO_2 nanoclusters synthesized by microwave processing. *RSC Adv.* **2016**, *6*, 9663–9671. [CrossRef]
63. Rabea, E.I.; Badawy, M.E.-T.; Stevens, C.V.; Smagghe, G.; Steurbaut, W. Chitosan as antimicrobial agent: Applications and mode of action. *Biomacromolecules* **2003**, *4*, 1457–1465. [CrossRef] [PubMed]
64. Qian, T.; Su, H.; Tan, T. The bactericidal and mildew-proof activity of a TiO_2–chitosan composite. *J. Photochem. Photobiol. A Chem.* **2011**, *218*, 130–136. [CrossRef]
65. Campoli-Richards, D.M.; Monk, J.P.; Price, A.; Benfield, P.; Todd, P.A.; Ward, A. Ciprofloxacin. *Drugs* **1988**, *35*, 373–447. [CrossRef] [PubMed]

Article

Extraction–Pyrolytic Method for TiO$_2$ Polymorphs Production

Vera Serga [1,2], Regina Burve [1,2], Aija Krumina [1], Marina Romanova [3], Eugene A. Kotomin [2] and Anatoli I. Popov [2,*]

[1] Institute of Inorganic Chemistry, Faculty of Materials Science and Applied Chemistry, Riga Technical University, Paula Valdena 3/7, LV-1048 Riga, Latvia; vera.serga@rtu.lv (V.S.); regina.burve@rtu.lv (R.B.); aija.krumina_4@rtu.lv (A.K.)
[2] Institute of Solid-State Physics, University of Latvia, Kengaraga 8, LV-1063 Riga, Latvia; kotomin@latnet.lv
[3] Institute of Biomedical Engineering and Nanotechnologies, Riga Technical University, Viskalu 36A, LV-1006 Riga, Latvia; marina.romanova@rtu.lv
* Correspondence: popov@latnet.lv

Citation: Serga, V.; Burve, R.; Krumina, A.; Romanova, M.; Kotomin, E.A.; Popov, A.I. Extraction–Pyrolytic Method for TiO$_2$ Polymorphs Production. *Crystals* 2021, 11, 431. https://doi.org/10.3390/cryst11040431

Academic Editors: Anton Meden, Philip Lightfoot and Pier Carlo Ricci

Received: 17 February 2021
Accepted: 14 April 2021
Published: 16 April 2021

Publisher's Note: MDPI stays neutral with regard to jurisdictional claims in published maps and institutional affiliations.

Copyright: © 2021 by the authors. Licensee MDPI, Basel, Switzerland. This article is an open access article distributed under the terms and conditions of the Creative Commons Attribution (CC BY) license (https://creativecommons.org/licenses/by/4.0/).

Abstract: The unique properties and numerous applications of nanocrystalline titanium dioxide (TiO$_2$) are stimulating research on improving the existing and developing new titanium dioxide synthesis methods. In this work, we demonstrate for the first time the possibilities of the extraction–pyrolytic method (EPM) for the production of nanocrystalline TiO$_2$ powders. A titanium-containing precursor (extract) was prepared by liquid–liquid extraction using valeric acid C$_4$H$_9$COOH without diluent as an extractant. Simultaneous thermogravimetric analysis and differential scanning calorimetry (TGA–DSC), as well as the Fourier-transform infrared (FTIR) spectroscopy were used to determine the temperature conditions to fabricate TiO$_2$ powders free of organic impurities. The produced materials were also characterized by X-ray diffraction (XRD) analysis and transmission electron microscopy (TEM). The results showed the possibility of the fabrication of storage-stable liquid titanium (IV)-containing precursor, which provided nanocrystalline TiO$_2$ powders. It was established that the EPM permits the production of both monophase (anatase polymorph or rutile polymorph) and biphase (mixed anatase–rutile polymorphs), impurity-free nanocrystalline TiO$_2$ powders. For comparison, TiO$_2$ powders were also produced by the precipitation method. The results presented in this study could serve as a solid basis for further developing the EPM for the cheap and simple production of nanocrystalline TiO$_2$-based materials in the form of doped nanocrystalline powders, thin films, and composite materials.

Keywords: titanium dioxide; anatase; rutile; polymorphs; extraction–pyrolytic method

1. Introduction

Among many functional nanomaterials, nanocrystalline titanium dioxide (TiO$_2$) powders are of great interest due to their unique properties and numerous practical applications [1–11].

The current interest in titanium dioxide-based nanostructured materials is primarily associated with their high-tech applications: solar cells (dye-sensitized, quantum dots-sensitized and perovskite), lithium-ion batteries, supercapacitors, gas sensors and, etc. [1–5,12]. Moreover, active investigations are related to the photocatalytic activity of TiO$_2$-based materials, including nanopowders and thin films. Due to chemical stability, non-toxicity, low cost, and high availability, titanium dioxide is considered the most promising photocatalyst for the degradation of organic pollutants in water and air, as well as for water splitting and hydrogen production [1–3,7,8,13–19]. However, TiO$_2$ is a wide bandgap semiconductor (3.2 and 3.02 eV for the anatase and rutile phases, respectively [20]) that requires UV light (5% in the solar spectrum) for its activation. To reduce the bandgap, TiO$_2$ should be either doped (e.g., with N, Ta) or used in the form of nanotubes [13,21–26]. Other important studies are related to the applications of TiO$_2$ as protective coatings in microelectronic and optical devices and as luminescent compounds [27–32].

TiO$_2$ forms three naturally occurring polymorphic crystalline modifications in the form of the corresponding minerals: brookite with rhombic, anatase and rutile with a tetragonal crystal lattice [1,4,33]. Rutile is the most thermodynamically stable modification. During heating, anatase and brookite irreversibly transform into rutile, and the stability of the crystalline modification depends on the size of its constituent crystallites [34,35]. Both the temperature of phase transformation and the properties of the produced nanostructured materials are largely determined by their manufacturing technology [36].

Highly dispersed titanium dioxide-based materials for various applications on a laboratory scale are produced by such well-known wet chemistry methods as sol–gel, microemulsion, precipitation, hydrothermal, solvothermal, electrochemical, sonochemical and microwave [2–5,9–11,21,22,25,26,37–42]. These methods allow fabricating TiO$_2$ nanostructures with different phase compositions and morphology, in particular as nanoparticles, nanorods, nanowires, nanotubes and mesoporous structures. The most promising and widely used method for producing TiO$_2$ is the sol–gel method [3–5,8,9,22,37,41,43], allowing obtaining TiO$_2$ powders with well-defined particle size and shape, excellent purity and homogeneity [37,43]. In the framework of the mentioned methods, inorganic salts (e.g., titanium tetrachloride TiCl$_4$) or organometallic compounds, such as metal alkoxides (e.g., titanium (IV) isopropoxide Ti[OCH(CH$_3$)$_2$]$_4$) are usually used as titanium-containing precursors. However, these compounds have high reactivity with water, which must be taken into account both during the material synthesis to ensure good reproducibility and during the follow-up storage. It should be mentioned that titanium alkoxides are expensive and not environmentally friendly.

Thus, to date, there is a huge number of publications presenting various methods for synthesizing highly dispersed titanium dioxide-based materials with a wide range of functional properties. Nevertheless, the current pace of technological development requires new synthesis approaches characterized by simplicity, ease of scaling, good reproducibility, use of inexpensive raw materials, and allowing the production of materials with the required characteristics. The extraction–pyrolytic method (EPM) could be considered as one of these new developments.

The EPM is used to fabricate homogeneous nanocrystalline powders and films of oxide materials for various purposes [44–48]. The EPM belongs to wet chemistry methods. Using the EPM, the following steps are required: fabrication of extract (metal-containing precursor) via the method of exchange extraction by fatty (aliphatic monocarboxylic straight- or branched-chain) acids with the addition of alkali [49] and following thermal treatment—pyrolysis. This technique is quite simple, inexpensive and does not require complex equipment. One of the important advantages of the EPM is using organic extracts (solutions of metal carboxylates in a carboxylic acid or solvent) as metal-containing precursors. Such precursors are resistant to humidity and do not crystallize during long-term storage. In addition, high-purity inorganic metal salts are not required for their preparation. During liquid extraction, the target component is purified from impurities. The liquid extraction of metal ions by monocarboxylic acid (HR) proceeds via a cation exchange mechanism and can be generally represented by Equation (1):

$$Me^{n+}_{(w)} + nHR_{(o)} \leftrightarrow MeR_{n(o)} + nH^+_{(w)} \qquad (1)$$

where the subscripts w and o denote the aqueous and organic phases, respectively.

Alkali is added to the extraction system to increase the efficiency of target metal extraction since monocarboxylic acids themselves (with or without a diluent) are usually ineffective extractants [49].

To date, the EPM has already been applied for producing photoactive titanium dioxide films [45]. As the initial components for preparing the Ti-containing extract, the authors used an aqueous solution of titanium (IV) oxysulfate TiOSO$_4$ and α-branched monocarboxylic acids of C$_5$–C$_9$ fractions as an extractant.

The aim of this work is to develop the EPM for the production of nanocrystalline TiO$_2$ powders using valeric acid-based extracts; and to study the effect of pyrolysis conditions on the phase composition, the mean crystallite size, and morphology of the fabricated materials. In addition, the results acquired by the EPM are compared with those related to the simplest and widely known production method—precipitation. In both approaches, the initial components are a freshly prepared aqueous solution of titanium (III) chloride as a titanium source and an aqueous solution of sodium hydroxide as an alkaline agent.

2. Materials and Methods

2.1. Preparation of the Precursors

2.1.1. Preparation of Aqueous Solution of Titanium (III) Chloride $TiCl_3$

An aqueous solution of $TiCl_3$ in diluted hydrochloric acid HCl with a metal concentration of 0.1 M was used as a titanium source. It was prepared immediately before both extraction and precipitation. For this, 1.200 g of titanium powder (particle size d = 63–100 µm) was dissolved in 60 mL of HCl solution (1:1) during heating until the metal was completely dissolved. Thereafter, the solution was cooled down and diluted with distilled water to a volume of 250 mL.

2.1.2. Preparation of Titanium-Containing Precursors (E) via Liquid–Liquid Extraction

Valeric acid C_4H_9COOH without diluent was used as an extractant. During preparing the precursor E1, the initial ratio of the volumes of the aqueous (V_w) and organic (V_o) phases in the extraction system was 3:1. For the extraction, the extractant and $TiCl_3$ solution (pH ~0.65) were placed in a separatory funnel, and 1 M NaOH solution was added step-by-step. When the organic phase (extract) turned deep blue, the addition of alkali was stopped. After a clear phase separation (~10 min), the aqueous phase was removed from the funnel, and its pH value was around 1.15. The organic phase was filtered through a cotton filter to remove water droplets.

To increase the titanium content in the organic phase for preparing the precursor E2, the initial $V_w:V_o$ ratio was taken as 5:1. The metal was extracted from $TiCl_3$ solution with a pH value of ~0.74. Moreover, the addition of an alkaline solution was continued until a saturated solution of titanium valerate $Ti(C_4H_9COO)_3$ in valeric acid was obtained, i.e., a finely dispersed precipitate appeared in the organic phase. As a result, the achieved pH value of the aqueous phase after extraction was about 1.23. To separate a small amount of the formed precipitate and to obtain a true solution, the organic phase was filtered through a double thick paper filter.

2.1.3. Preparation of Titanium-Containing Precursor (P) via Precipitation

As the first step, alkaline hydrolysis of $TiCl_3$ solution was carried out at room temperature. 0.5 M NaOH solution was added dropwise (at a rate of ~ 3 mL/min) under vigorous stirring until the pH of the aqueous phase reached ~6.0. Then, the mixture was left to stay for a day. This was followed by filtration, multiply washing of the resulting precipitate with distilled water (the presence of chloride ions in the decanted solution was controlled with an $AgNO_3$ solution) and, after all, with ethanol. The precipitate was dried at room temperature for 36 hours, ground in an agate mortar and used as a precursor (P).

2.2. Thermal Treatment of Precursors

The resulting precursors E1 and E2, as solutions, and precursor P as powder were heated from room temperature to 350–750 °C at a heating rate of 10°/min, annealed for an hour and rapidly cooled down under ambient conditions. Such thermal treatment was performed in laboratory furnace SNOL 8.2/1100. Thereafter, the produced samples were ground by pestle in an agate mortar and collected. For further investigations, only as-prepared powders without any additional posttreatment were used.

2.3. Characterization Methods

The metal concentration in the resulting precursors E was determined by the gravimetric method [50].

The thermal behavior of all the produced precursors was studied by simultaneous thermogravimetric analysis and differential scanning calorimetry (TGA–DSC) using the STA PT1600 (LINSEIS). The samples under test were heated from room temperature to 700 °C or 1000 °C at a rate of 10°/min in the static air atmosphere.

The phase composition of the produced materials was investigated by the X-ray diffraction (XRD) method (diffractometer D8 Advance, Bruker Corporation) with CuKα radiation (λ = 1.5418 Å). The XRD patterns were referenced to the PDF ICDD 00-021-1272 for anatase phase of TiO_2, PDF ICDD 00-021-1276 for rutile phase of TiO_2, and PDF ICDD 00-014-0277 for sodium polytitanate ($Na_2Ti_6O_{13}$) identification. The mean crystallite size (d) of the titanium dioxide was defined from the half-width of the diffraction peaks (101) of anatase (d_A) and (110) one of rutile (d_R) by the Scherrer method (EVA software). The weight fraction of the rutile phase (W_R) was determined by Gribb and Banfield [34] using integrated intensities A (areas) of the most intense diffractions peaks as follows (Equation (2)).

$$W_R = \frac{A_R}{0.884 A_A + A_R} \cdot 100\% \qquad (2)$$

IR spectra were recorded at room temperature using Bruker Tensor II FTIR spectrometer at a resolution of 4 cm^{-1} and 36 scans for each spectrum. TiO_2 powder was mixed with KBr, and the pellets with a 7 mm diameter were prepared using Specac Mini-Pellet press under a load of 2000 kg. The morphology of the samples was examined by transmission electron microscopy (TEM) (FEI Tecnai G2 F20 operating at 200 kV).

3. Results

3.1. Precursors Characterization

Titanium-containing precursors (E) During extraction, Ti^{3+} cations are transferred from the aqueous phase into the organic phase as $Ti(C_4H_9COO)_3$, and the organic solution gradually turns deep blue. As a result of the storage of the produced titanium-containing extract E1 in a glass flask, the organic solution underwent discoloration, first, gradually and after ~60 minutes complete. The process was likely associated with the oxidation of Ti (III) to Ti (IV) by atmospheric oxygen. To our knowledge, there is no data on the composition of the final titanium (IV) carboxylate formed this way in an organic solution. However, it can be assumed that this compound may have the following composition: $Ti(C_4H_9COO)_4$ and/or $(C_4H_9COO)_3TiOTi(OOCC_4H_9)_3$.

Note that the discoloration of the organic titanium-containing extract E2 was observed already during filtration. According to the results of the gravimetric analysis, the titanium concentration in the precursors E1 and E2 was 0.14 M and 0.50 M, respectively.

Thus, colorless transparent organic solutions with different titanium concentrations were prepared. Upon storage of E1 and E2 precursors in glass flasks with ground-glass stoppers at room temperature, no changes in color and transparency (homogeneity) were observed.

Titanium-containing precursor (P) As a result of the produced precipitate (gel) storage during the day, its color changed from deep gray-blue to white because of the oxidation of titanium (III) hydroxide by atmospheric oxygen and the formation of hydrated titanium dioxide (titanium oxyhydrate) $TiO_2·nH_2O$ [51].

3.2. Thermal Behavior of Precursors E1, E2, and P

The main thermal decomposition products of the salts of many carboxylic acids are ketones and the corresponding metal oxides, while the temperature of their decomposition is characteristic for each certain compound [52]. This is why the study of the thermal behavior of the extracts produced during the EPM (solutions of metal carboxylates in carboxylic acid or diluent) is rather important and necessary for determining the minimal pyrolysis temperature for the production of organic impurity-free oxide materials.

The results of TG-DSC analysis of liquid precursors E1 and E2 with different titanium concentrations are shown in Figure 1. According to the data presented, studied precursors demonstrate similar thermal behavior during the heating process. At the same time, the thermal effects are more pronounced for the precursor with higher titanium concentration (E2) and just these results (see Figure 1B) will be discussed in detail.

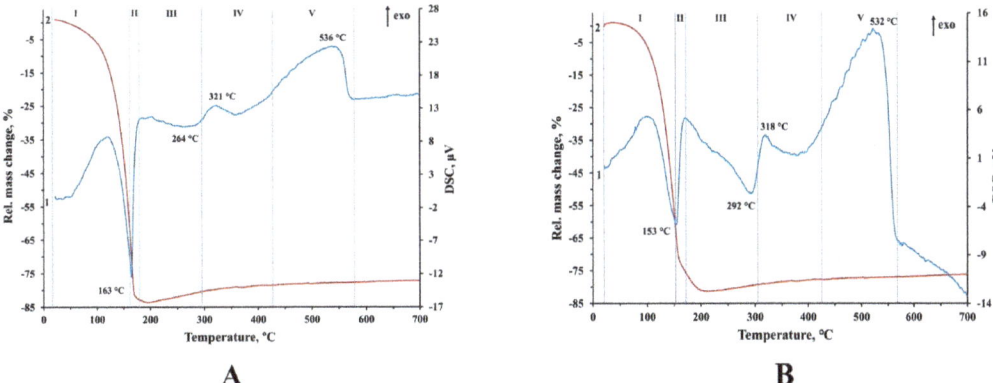

Figure 1. DSC (1) and TGA (2) curves of the precursors produced by extraction–pyrolytic method (EPM): (**A**) E1; (**B**) E2.

The precursor E2 is thermally stable up to a temperature of ~30 °C. The first endothermic peak on the DSC curve at ~153 °C is accompanied by active sample weight loss. In the temperature range from ~30 °C to ~153 °C (fragment I), this weight loss is mainly associated with the removal of free extractant (valeric acid) and co-extracted water, while at a further temperature increase (fragment II)—with the decomposition of the titanium (IV) carboxylate. The second broad asymmetric endothermic peak is observed at ~180–306 °C (fragment III). In the region of this peak, the decomposition of titanium carboxylate still continues and is followed by active evaporation of the organic decomposition product (probably, dibutyl ketone $C_4H_9COC_4H_9$ with $T_{boiling}$ = 182–187 °C). According to the TG curve, the weight loss reaches ~82% at ~210 °C and stops. A further temperature rise to 700 °C is accompanied by a gradual increase in the sample weight by ~4%. That is most likely associated with the gradual oxidation of titanium monoxide TiO to TiO_2 using the proposed in Ref. [53] decomposition mechanism of the metal (IV) carboxylate via the formation of metal monoxide as intermediate. Moreover, at ~210–300 °C, this process occurs simultaneously with the removal of volatile organic decomposition products. The increase in sample weight observed on the TG curve (Figure 1B, curve 2) shows that TiO oxidation is the dominant process. At the same time, the predominance of the endothermic evaporation process is observed as well (see curve 1 in Figure 1B). A weak exothermic peak at ~318 °C (fragment IV) on the DSC curve is caused by the combustion of gaseous organic residue. In the temperature range, ~433–561 °C (fragment V), an intense asymmetric exothermic peak assumes the superposition of several thermal effects: crystallization of an amorphous phase, anatase-to-rutile polymorphic transformation and pyrocarbon burnout. Thus, according to the analysis of the obtained results, it could be assumed that upon heating, the organic decomposition product (most likely, ketone) is removed after forming TiO and its oxidation to TiO_2.

For comparison, the thermal behavior of a solid precursor P (titanium oxyhydrate sample) was also studied. The presented thermogram (Figure 2, curve 1) shows two endothermic and one exothermic peak. The endothermic effect observed at ~56–150 °C (fragment I) is accompanied by active weight loss (~20%) of the dried precipitate due to the removal of adsorbed water. With a further increase in temperature, the sample loses crystallization water. This process is accompanied by a wide endothermic peak at ~208–308 °C (fragment II) and a small weight loss (~5%). The intense exothermic peak at ~776 °C (fragment III) is associated with the crystallization of titanium dioxide and polymorphic anatase-to-rutile transformation. Ongoing slight loss of sample weight is probably due to the continuation of the dehydration process.

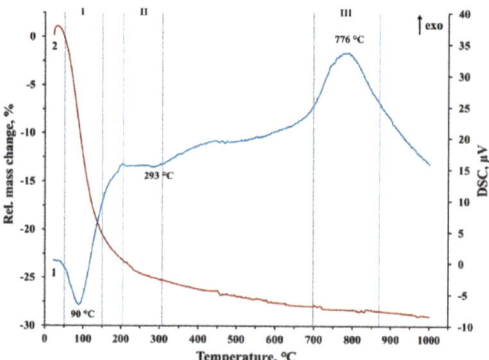

Figure 2. DSC (1) and TG (2) curves of the precursor (P).

According to Figures 1 and 2, due to the different chemical compositions, the thermal behavior of the studied precursors differs significantly. Thus, the observed weight loss of the precursor P upon heating is associated with successive dehydration processes. At the same time, thermal transformations in precursors E are associated with the complex decomposition of titanium carboxylate, which is preceded by the evaporation processes of the solvent (valeric acid) and co-extracted water being the parts of the extracts.

3.3. XRD Analysis

To obtain a solid final product from precursors E1 and E2, based on the TG-DSC results (see Figure 1), the minimal temperature of pyrolysis was chosen as 350 °C. To study the effect of the precursor preparation method on the phase composition, anatase-to-rutile transformation temperature, and the mean crystallite size of TiO_2, heat treatment of precursors E and P was carried out in the range from 450 °C to 750 °C with a temperature step of 100 °C. Table 1 summarizes the results of the XRD analysis of all the produced materials.

The study of the regularities of phase formation during the pyrolysis of the precursor E1 testifies (Figure 3, Table 1, samples E1-1–E1-6) that amorphous powders are produced at temperatures of 350 °C and 400 °C. The crystallization of anatase polymorph begins at 450 °C, while the polymorphic anatase-to-rutile transformation starts at 650 °C. TiO_2 powder produced at 750 °C contains rutile polymorph with only a small anatase admixture ($W_A = 1.1\%$).

According to the XRD analysis (Figure 4, Table 1, samples E2-1–E2-6), the heat treatment of a more concentrated precursor E2 at 400 °C corresponds to the beginning of the anatase phase crystallization. Pyrolysis of the precursor at 550 °C and 650 °C leads to the gradual polymorphic transformation of anatase to rutile with a simultaneous increase in the mean crystallite size of anatase from ~20 nm to ~35 nm and of rutile from ~30 nm to ~45 nm, respectively. As a result of heat treatment at 750 °C, a monophase product consisting of a rutile polymorph with d_R ~53 nm is formed.

Thus, an increase in titanium concentration in the precursor solution from 0.14 M to 0.50 M decreases the temperature of anatase-to-rutile transformation by ~100 °C (see Figures 3 and 4, Table 1).

Table 1. Impact of the heat treatment conditions of titanium-containing precursors on the phase composition and mean crystallite size of the final products.

Sample Nr.	Production Conditions		XRD Analysis Results		
	Precursor	Pyrolysis Temperature T, °C	Phase Composition	d, nm	W, %
E1-1	E1	350	Amorphous	-	-
E2-1	E2		Amorphous	-	-
E1-2	E1	400	Amorphous	-	-
E2-2	E2		Anatase	5	100
E1-3	E1	450	Anatase	8	100
E2-3	E2		Anatase	9	100
P-1	P		Anatase	9	100
E1-4	E1	550	Anatase	15	100
E2-4	E2		Anatase / Rutile	20 / ~30	87.7 / 12.3
P2	P		Anatase	10	100
E1-5	E1	650	Anatase / Rutile	30 / ~40	80.9 / 19.1
E2-5	E2		Anatase / Rutile	~35 / 45	20.6 / 79.4
P3	P		Anatase / Rutile	14 / Discerned	96.4 / 3.6
E1-6	E1	750	Anatase / Rutile	Discerned / 65	1.1 / 98.9
E2-6	E2		Rutile	53	100
P-4	P		Rutile / $Na_2Ti_6O_{13}$	68 / -	100 / -

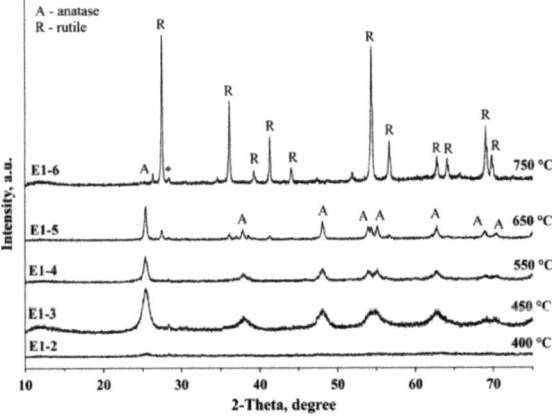

Figure 3. XRD patterns of nanocrystalline TiO_2 powders produced from the precursor E1 at different pyrolysis temperatures. * Signal of a silicon substrate from the measuring cuvette.

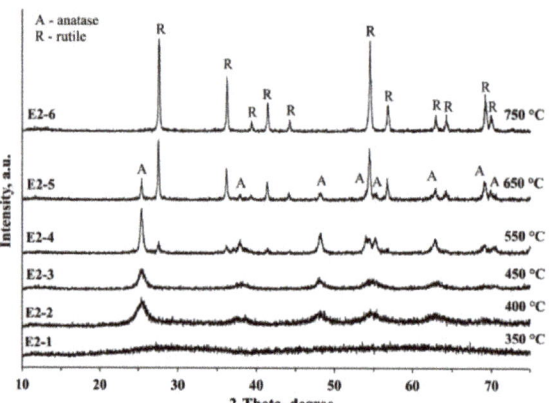

Figure 4. X-ray diffraction patterns of nanocrystalline TiO_2 powders produced from the precursor E2 at different pyrolysis temperatures.

According to XRD analysis, precursor P is amorphous (Figure 5, Table 1). The anatase phase, produced as a result of the heat treatment of precursor P, is similar to these at the precursor E1 treatment at 450 °C and 550 °C. The increase of the processing temperature to 650 °C or 750 °C leads to the formation of two phases of TiO_2. Moreover, depending on the heat treatment temperature, either anatase or rutile is a dominating phase (Figure 5, samples P-3 and P-4). It was also found two processes occur simultaneously at 750 °C: the polymorphic transformation of anatase into rutile and the crystallization of the admixture phase, $Na_2Ti_6O_{13}$. This phase is a product of the interaction of NaOH with TiO_2 at high temperatures, i.e., during the preparation of a precursor P, it is impossible to completely remove the residual amounts of NaOH by washing the precipitate (gel). In the case of the EPM, a system of two immiscible liquids is used, and the target product (titanium carboxylate) is dissolved in the organic phase, while water-soluble reaction components, in the aqueous phase. Hence, the presence of impurity phases in the TiO_2 samples was not established (see Figures 3 and 4).

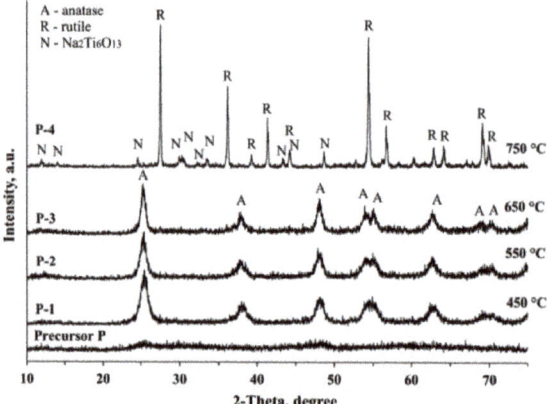

Figure 5. XRD patterns of nanocrystalline TiO_2 powders produced from the precursor P at different processing temperatures.

3.4. FTIR Spectroscopy

FTIR spectroscopy was used to determine the conditions for the thermal treatment of the precursor E2 that ensure complete removal of the organic component during TiO_2 production.

The FTIR spectra (see Figure 6) contain the peaks at 3449 cm^{-1} and 1622 cm^{-1}, which correspond to the stretching and bending vibrations of –OH groups. Weak absorption bands at 2362 cm^{-1} and 2332 cm^{-1} in the samples are associated with the presence of carbon dioxide CO_2 absorbed from the atmosphere [54]. In the case of the samples produced at 350 °C or 400 °C (Figure 6, samples E2-1 and E2-2), the spectra contain the absorption bands peaked at 1520 cm^{-1} and 1375 cm^{-1}, which indicate the presence of undecomposed organic residue in these materials [55,56]. The presence of TiO_2 in the studied materials is confirmed by a wide absorption band at ~1000 cm^{-1}–400 cm^{-1} associated with the vibrations of Ti–O–Ti bonds in the TiO_2 lattice [57,58]. In the mentioned spectral region, a shift of the maximum from 515 cm^{-1} to 442 cm^{-1} is observed upon the decrease in the precursor pyrolysis temperature from 550 °C to 350 °C (Figure 6, samples E2-4–E2-1). This fact may be related to the changes in the size of the produced TiO_2 particles, as described earlier in [41,59]. This is also consistent with the results of our XRD analysis (Table 1), under which a decrease in the pyrolysis temperature of the precursor E2 in this temperature range leads to a decrease in the mean crystallite size of anatase from 20 nm to 5 nm, and, finally, to amorphization. Thus, according to our results, to produce organic impurity-free TiO_2 powders via the EPM, the minimal pyrolysis temperature of the extracts (precursors E) should be 450 °C. The data obtained do not contradict the results of the TG-DSC analysis (Figure 1) presented above. A similar picture was observed at the comparison of the infrared spectra for bulk and nanosized AlN and $LaPO_4$ [60–62].

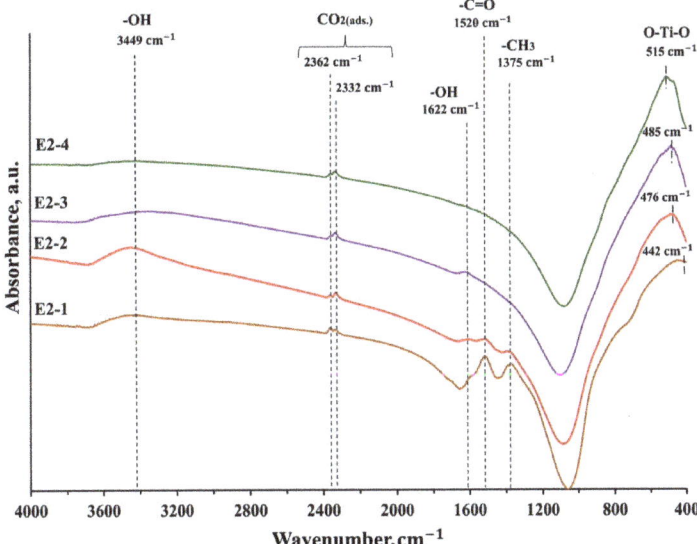

Figure 6. FTIR spectra of the samples produced from the precursor E2 at different pyrolysis temperatures: E2-1—350 °C, E2-2—400 °C, E2-3—450 °C, E2-4—550 °C.

3.5. Transmission Electron Microscopy

Figure 7 demonstrates TEM results for the anatase and rutile powders produced by the EPM and, for comparison, for the anatase sample produced by the precipitation method. According to the results obtained, the particles with irregular rounded shapes are formed as a result of the low-temperature treatment (450°C) of both precursors (Figure 7A,C).

Nanoparticles with a mean size of ~8 nm can be observed that is in line with the XRD results (d_A = 9 nm, Table 1).

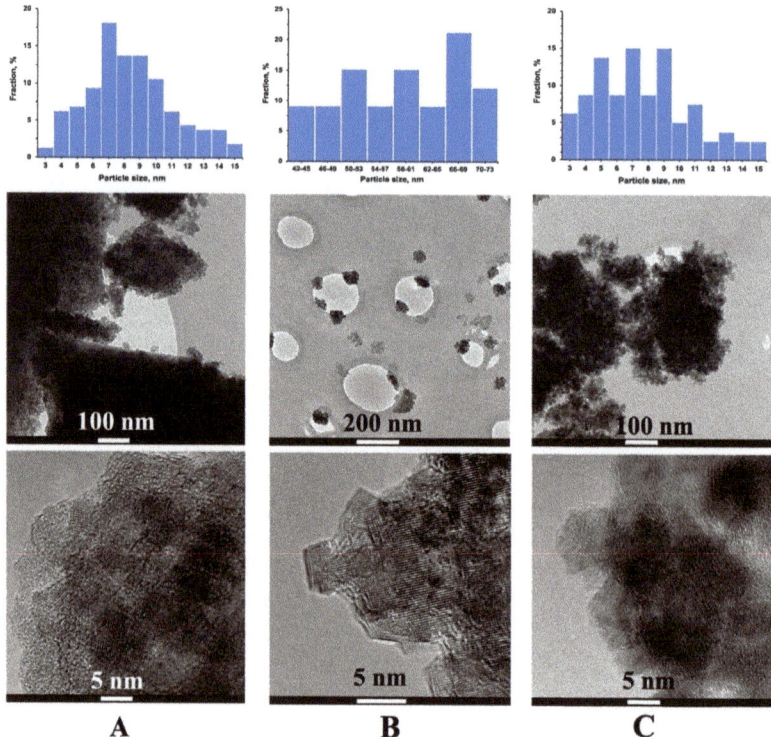

Figure 7. HR-TEM (a bottom raw), TEM (a medium raw) images and histograms of the particle size distribution (a top raw) of samples produced by the EPM (**A**,**B**) and precipitation method (**C**) at temperatures: (**A**)—450 °C (sample E2-3); (**B**)—750 °C (sample E2-6); (**C**)—450 °C (sample P-1).

In the case of the EPM, the increase in the pyrolysis temperature up to 750 °C leads to the formation of layered aggregates consisting of the faceted particles with a mean size of ~11 nm (Figure 7B). It is possible that the formation of such structures is associated with the thermal behavior of the precursor upon heating (see Section 3.2), in particular, with the effect of the pyrolysis products of the organic precursor on the nanoparticle surface. The average size of the aggregates is about 58 nm that is consistent with the XRD data (d_R = 53 nm, Table 1).

4. Conclusions

This study suggests an original two-stage approach for synthesizing nanocrystalline TiO$_2$ powders—the extraction–pyrolytic method (EPM).

The conditions for preparing titanium-containing extracts (precursors) using valeric acid without a diluent as an extractant were determined. The minimum temperature of pyrolysis (450 °C) of the precursors for organic impurity-free nanocrystalline TiO$_2$ production was established. We have shown that the phase composition of the resulting powders is affected by the pyrolysis temperature and titanium concentration in the precursor solution. According to the XRD results, depending on pyrolysis conditions, the produced TiO$_2$ samples contain anatase (d_A ~8–15 nm), mixed anatase-rutile or rutile (d_R ~53 nm) polymorphs. We have shown that the decrease in titanium concentration in the precursor solution from 0.50 to 0.14 M leads to the increase of the temperature of anatase-to-rutile polymorphic transformation by ~100 °C. Comparative analysis of the results for the materials produced by two methods—the EPM and the precipitation, revealed some differences. According to the XRD data, as a result of the heat treatment at 750 °C, impurity phases were not detected in the EPM-produced materials, while the $Na_2Ti_6O_{13}$ impurity phase was identified in the material produced by the precipitation method.

The results presented in this study could serve as a solid basis for further developing the EPM for the cheap and simple production of nanocrystalline TiO_2-based materials in the form of doped nanocrystalline powders, thin films and composite materials.

Author Contributions: Conceptualization, V.S. and A.I.P.; methodology, V.S. and R.B.; software, A.K. and M.R.; formal analysis, R.B., A.K. and M.R.; investigation, V.S. and R.B.; resources, E.A.K. and A.I.P.; data curation, E.A.K.; writing—original draft preparation, V.S., R.B. and A.I.P.; writing—review and editing, E.A.K. and A.I.P.; visualization, A.K. and M.R.; supervision, V.S. and A.I.P.; project administration, E.A.K.; funding acquisition, E.A.K. All authors have read and agreed to the published version of the manuscript.

Funding: This study was partly supported by the M-ERA.NET project SunToChem.

Institutional Review Board Statement: Not applicable.

Informed Consent Statement: Not applicable.

Data Availability Statement: Not applicable.

Acknowledgments: The authors thank V. Kuzovkov, A. Lushchik and M. Lushchik for many useful discussions. The research was (partly) performed in the Institute of Solid State Physics, University of Latvia ISSP UL. ISSP UL as the Center of Excellence is supported through the Framework Program for European universities Union Horizon 2020, H2020-WIDESPREAD-01–2016–2017-TeamingPhase2 under Grant Agreement No. 739508, CAMART2 project.

Conflicts of Interest: The authors declare no conflict of interest.

References

1. Gupta, S.M.; Tripathi, M. A review of TiO_2 nanoparticles. *Sci. Bull.* **2011**, *56*, 1639–1657. [CrossRef]
2. Ge, M.; Cao, C.; Huang, J.; Li, S.; Chen, Z.; Zhang, K.-Q.; Al-Deyab, S.S.; Lai, Y. A review of one-dimensional TiO_2 nanostructured materials for environmental and energy applications. *J. Mater. Chem. A* **2016**, *4*, 6772–6801. [CrossRef]
3. Macwan, D.P.; Dave, P.N.; Chaturvedi, S. A review on nano-TiO_2 sol-gel type synthesis and its applications. *J. Mater. Sci.* **2011**, *46*, 3669–3686. [CrossRef]
4. Dubey, R.S.; Krishnamurthy, K.V.; Singh, S. Experimental studies of TiO_2 nanoparticles synthesized by sol-gel and solvothermal routes for DSSCs application. *Results Phys.* **2019**, *14*, 102390. [CrossRef]
5. Singh, R.; Ryu, I.; Yadav, H.; Park, J.; Jo, J.W.; Yim, S.; Lee, J.-J. Non-hydrolytic sol-gel route to synthesize TiO_2 nanoparticles under ambient condition for highly efficient and stable perovskite solar cells. *Sol. Energy* **2019**, *185*, 307–314. [CrossRef]
6. Lingaraju, K.; Basavaraj, R.B.; Jayanna, K.; Bhavana, S.; Devaraja, S.; Kumar Swamy, H.M.; Nagaraju, G.; Nagabhushan, H.; Raja Naika, H. Biocompatible fabrication of TiO_2 nanoparticles: Antimicrobial, anticoagulant, antiplatelet, direct hemolytic and cytotoxicity properties. *Inorg. Chem. Commun.* **2021**, *127*, 10850. [CrossRef]
7. Chen, D.; Cheng, Y.; Zhou, N.; Chen, P.; Wang, Y.; Li, K.; Huo, S.; Cheng, P.; Peng, P.; Zhang, R.; et al. Photocatalytic degradation of organic pollutants using TiO_2-based photocatalysts: A review. *J. Clean. Prod.* **2020**, *268*, 121725. [CrossRef]
8. Haider, A.J.; AL-Anbari, R.H.; Kadhim, G.R.; Salame, C.T. Exploring potential environmental applications of TiO_2 nanoparticles. *Energy Procedia* **2017**, *119*, 332–345. [CrossRef]
9. Lusvardi, G.; Barani, C.; Giubertoni, F.; Paganelli, G. Synthesis and characterization of TiO_2 nanoparticles for the reduction of water pollutants. *Materials* **2017**, *10*, 1208. [CrossRef] [PubMed]
10. Chen, P.C.; Chen, C.C.; Chen, S.H. A review on production, characterization, and photocatalytic applications of TiO_2 nanoparticles and nanotubes. *Curr. Nanosci.* **2017**, *13*, 373–393. [CrossRef]
11. Wang, S.; Yu, H.; Yuan, S.; Zhao, Y.; Wang, Z.; Fang, J.; Zhang, M.; Shi, L. Synthesis of triphasic, biphasic, and monophasic TiO_2 nanocrystals and their photocatalytic degradation mechanisms. *Res. Chem. Intermed.* **2016**, *42*, 3775–3788. [CrossRef]
12. Ramanavicius, S.; Ramanavicius, A. Insights in the application of stoichiometric and non-stoichiometric titanium oxides for the design of sensors for the determination of gases and VOCs (TiO_{2-x} and $TinO_{2n-1}$ vs. TiO_2). *Sensors* **2020**, *20*, 6833. [CrossRef] [PubMed]
13. Zhukovskii, Y.F.; Piskunov, S.; Lisovski, O.; Bocharov, D.; Evarestov, R.A. Doped 1D nanostructures of transition-metal oxides: First-principles evaluation of photocatalytic suitability. *Isr. J. Chem.* **2017**, *57*, 461–476. [CrossRef]

14. Sidaraviciute, R.; Kavaliunas, V.; Puodziukynas, L.; Guobiene, A.; Martuzevicius, D.; Andrulevicius, M. Enhancement of photocatalytic pollutant decomposition efficiency of surface mounted TiO_2 via lithographic surface patterning. *Environ. Technol. Innov.* **2020**, *19*, 100983. [CrossRef]
15. Nosaka, Y.; Nosaka, A.Y. Generation and detection of reactive oxygen species in photocatalysis. *Chem. Rev.* **2017**, *117*, 11302–11336. [CrossRef] [PubMed]
16. Tamm, A.; Seinberg, L.; Kozlova, J.; Link, J.; Pikma, P.; Stern, R.; Kukli, K. Quasicubic α-Fe_2O_3 nanoparticles embedded in TiO_2 thin films grown by atomic layer deposition. *Thin Solid Films* **2016**, *612*, 445–449. [CrossRef]
17. Rempel, A.A.; Kuznetsova, Y.V.; Dorosheva, I.B.; Valeeva, A.A.; Weinstein, I.A.; Kozlova, E.A.; Saraev, A.A.; Selishchev, D.S. High Photocatalytic Activity Under Visible Light of Sandwich Structures Based on Anodic TiO_2/CdS Nanoparticles/Sol–Gel TiO_2. *Top. Catal.* **2020**, *63*, 130–138. [CrossRef]
18. Tuckute, S.; Varnagiris, S.; Urbonavicius, M.; Lelis, M.; Sakalauskaite, S. Tailoring of TiO_2 film crystal texture for higher photocatalysis efficiency. *Appl. Surf. Sci.* **2019**, *489*, 576–583. [CrossRef]
19. Kenmoe, S.; Lisovski, O.; Piskunov, S.; Bocharov, D.; Zhukovskii, Y.F.; Spohr, E. Water adsorption on clean and defective anatase TiO_2 (001) nanotube surfaces: A surface science approach. *J. Phys. Chem. B* **2018**, *122*, 5432–5440. [CrossRef]
20. Wunderlich, W.; Oekermann, T.; Miao, L.; Hue, N.T.; Tanemura, S.; Tanemura, M. Electronic properties of Nano-porous TiO_2- and ZnO- thin films—Comparison of simulations and experiments. *J. Ceram. Process. Res.* **2004**, *5*, 343–354.
21. Knoks, A.; Kleperis, J.; Grinberga, L. Raman spectral identification of phase distribution in anodic titanium dioxide coating. *Proc. Estonian Acad. Sci.* **2017**, *66*, 422–429. [CrossRef]
22. Brik, M.G.; Antic, Ž.M.; Vukovic, K.; Dramicanin, M.D. Judd-Ofelt analysis of Eu^{3+} emission in TiO_2 anatase nanoparticles. *Mater. Trans.* **2015**, *56*, 1416–1418. [CrossRef]
23. Nishioka, S.; Yanagisawa, K.; Lu, D.; Vequizo, J.J.M.; Yamakata, A.; Kimoto, K.; Inada, M.; Maeda, K. Enhanced water splitting through two-step photoexcitation by sunlight using tantalum/nitrogen-codoped rutile titania as a water oxidation photocatalyst. *Sustain. Energy Fuels* **2019**, *3*, 2337–2346. [CrossRef]
24. Kavaliunas, V.; Krugly, E.; Sriubas, M.; Mimura, H.; Laukaitis, G.; Hatanaka, Y. Influence of Mg, Cu, and Ni dopants on amorphous TiO_2 thin films photocatalytic activity. *Materials* **2020**, *13*, 886. [CrossRef] [PubMed]
25. Wu, F.; Hu, X.; Fan, J.; Sun, T.; Kang, L.; Hou, W.; Zhu, C.; Liu, H. Photocatalytic activity of Ag/TiO_2 nanotube arrays enhanced by surface plasmon resonance and application in hydrogen evolution by water splitting. *Plasmonics* **2013**, *8*, 501–508. [CrossRef]
26. Linitis, J.; Kalis, A.; Grinberga, L.; Kleperis, J. Photo-activity research of nano-structured TiO_2 layers. *IOP Conf. Ser. Mater. Sci. Eng.* **2011**, *23*, 012010. [CrossRef]
27. Kozlovskiy, A.; Shlimas, D.; Kenzhina, I.; Boretskiy, O.; Zdorovets, M. Study of the effect of low-energy irradiation with O^{2+} ions on radiation hardening and modification of the properties of thin TiO_2 films. *J. Inorg. Organomet. Polym. Mater.* **2021**, *31*, 790–801. [CrossRef]
28. Mattsson, M.S.M.; Azens, A.; Niklasson, G.A.; Granqvist, C.G.; Purans, J. Li intercalation in transparent Ti-Ce oxide films: Energetics and ion dynamics. *J. Appl. Phys.* **1997**, *81*, 6432–6437. [CrossRef]
29. Dukenbayev, K.; Kozlovskiy, A.; Kenzhina, I.; Berguzinov, A.; Zdorovets, M. Study of the effect of irradiation with Fe^{7+} ions on the structural properties of thin TiO_2 foils. *Mater. Res. Express* **2019**, *6*, 046309. [CrossRef]
30. Kiisk, V.; Akulits, K.; Kodu, M.; Avarmaa, T.; Mändar, H.; Kozlova, J.; Eltermann, M.; Puust, L.; Jaaniso, R. Oxygen-sensitive photoluminescence of rare earth ions in TiO_2 thin films. *J. Phys. Chem. C* **2019**, *123*, 17908–17914. [CrossRef]
31. Milovanov, Y.S.; Gavrilchenko, I.V.; Gayvoronsky, V.Y.; Kuznetsov, G.V.; Skryshevsky, V.A. Impact of Nanoporous Metal Oxide Morphology on Electron Transfer Processes in Ti–TiO_2–Si Heterostructures. *J. Nanoelectron. Optoelectron.* **2014**, *9*, 432–436. [CrossRef]
32. Reklaitis, I.; Radiunas, E.; Malinauskas, T.; Stanionytė, S.; Juška, G.; Ritasalo, R.; Pilvi, T.; Taeger, S.; Strassburg, M.; Tomašiūnas, R. A comparative study on atomic layer deposited oxide film morphology and their electrical breakdown. *Surf. Coat. Technol.* **2020**, *399*, 126123. [CrossRef]
33. Luchinsky, G.P. *Chemistry of the Titanium*; Khimija: Moskow, Russia, 1971. (In Russian)
34. Gribb, A.A.; Banfield, J.F. Particle size effects on transformation kinetics and phase stability in nanocrystalline TiO_2. *Amer. Miner.* **1997**, *82*, 717–728. [CrossRef]
35. Zhang, H.; Banfield, J.F. Thermodynamic analysis of phase stability of nanocrystalline titania. *J. Mater. Chem.* **1998**, *8*, 2073–2076. [CrossRef]
36. Hanaor, D.A.H.; Sorell, C.C. Review of the anatase to rutile phase transformation. *J. Mater. Sci.* **2011**, *46*, 855–874. [CrossRef]
37. Gupta, S.M.; Tripathi, M. A review on the synthesis of TiO_2 nanoparticles by solution route. *Cent. Eur. J. Chem.* **2012**, *10*, 279–294. [CrossRef]
38. Byranvand, M.M.; Kharat, A.N.; Fatholahi, L.; Beiranvand, Z.M. A review on synthesis of nano-TiO_2 via different methods. *J. Nanostruct.* **2013**, *3*, 1–9. [CrossRef]
39. Wang, Z.; Liu, S.; Cao, X.; Wu, S.; Liu, C.; Li, G.; Jiang, W.; Wang, H.; Wang, N.; Ding, W. Preparation and characterization of TiO_2 nanoparticles by two different precipitation methods. *Ceram. Int.* **2020**, *46*, 15333–15341. [CrossRef]
40. Wategaonkar, S.B.; Pawar, R.P.; Parale, V.G.; Nade, D.P.; Sargar, B.M.; Mane, R.K. Synthesis of rutile TiO_2 nanostructures by single step hydrothermal route and its characterization. *Mater. Today Proc.* **2020**, *23*, 444–451. [CrossRef]

41. Kusior, A.; Banas, J.; Trenczek-Zajac, A.; Zubrzycka, P.; Micek-Ilnicka, A.; Radecka, M. Structural properties of TiO_2 nanomaterials. *J. Mol. Struct.* **2018**, *1157*, 327–336. [CrossRef]
42. Sharma, A.; Karn, R.K.; Pandiyan, S.K. Synthesis of TiO_2 nanoparticles by sol-gel method and their characterization. *J. Basic Appl. Eng. Res.* **2014**, *1*, 1–5.
43. Toygun, S.; Konecoglu, G.; Kalpakli, Y. General principles of sol-gel. *J. Eng. Nat. Sci.* **2013**, *31*, 456–476.
44. Khol'kin, A.I.; Patrusheva, T.N. *Extraction-Pyrolytic Method: Fabrication of Functional Oxide Materials*; KomKniga: Moskow, Russian, 2006; ISBN 548-400-582-5. (In Russian)
45. Patrusheva, T.N.; Popov, V.S.; Prabhu, G.; Popov, A.V.; Ryzhenkov, A.V.; Snezhko, N.Y.; Morozchenko, D.A.; Zaikovskii, V.D.; Khol'kin, A.I. Preparation of a photoanode with a multilayer structure for solar cells by extraction pyrolysis. *Theor. Found. Chem. Eng.* **2014**, *48*, 454–460. [CrossRef]
46. Popov, A.I.; Shirmane, L.; Pankratov, V.; Lushchik, A.; Kotlov, A.; Serga, V.E.; Kulikova, L.D.; Chikvaidze, G.; Zimmermann, J. Comparative study of the luminescence properties of macro- and nanocrystalline MgO using synchrotron radiation. *Nucl. Instrum. Methods Phys. Res. B* **2013**, *310*, 23–26. [CrossRef]
47. Serga, V.; Burve, R.; Maiorov, M.; Krumina, A.; Skaudzius, R.; Zarkov, A.; Kareiva, A.; Popov, A. Impact of gadolinium on the structure and magnetic properties of nanocrystalline powders of iron oxides produced by the extraction-pyrolytic method. *Materials* **2020**, *13*, 4147. [CrossRef] [PubMed]
48. Burve, R.; Serga, V.; Krumina, A.; Poplausks, R. Preparation and characterization of nanocrystalline gadolinium oxide powders and films. *Key Eng. Mater.* **2020**, *850*, 267–272. [CrossRef]
49. Gindin, L.M. *Extraction Processes and Its Application*; Nauka: Moskow, Russia, 1984. (In Russian)
50. Sharlo, G. Quantitative Analysis of the Inorganic Compounds. In *Methods of the Analytical Chemistry*; Lur'e, Y.Y., Ed.; Himija: Moskow, Russia, 1969; Volume 2, ISBN 978-544-584-821-9. (In Russian)
51. Drozdov, A.A.; Zlomanov, G.N.; Mazo, G.N.; Spiridinov, F.M. Chemistry of the Transition Elements. In *Inorganic Chemistry*; Tretyakov, Y.D., Ed.; Akademija: Moskow, Russia, 2008; Volume 3, Part 1; pp. 56–99, ISBN 576-952-532-0. (In Russian)
52. Mehrotra, R.C.; Bohra, R. *Metal Carboxylates*; Academic Press: London, UK, 1983; ISBN 978-012-488-160-0.
53. Patil, K.C.; Chandrashekhar, G.V.; George, M.V.; Rao, C.N.R. Infrared spectra and thermal decompositions of metal acetates and dicarboxylates. *Can. J. Chem.* **1968**, *46*, 257–265. [CrossRef]
54. Smith, B.C. A process for successful infrared spectral interpretation. *Spectroscopy* **2016**, *31*, 14–21.
55. Stuart, B. Analytical techniques in the sciences. In *Infrared Spectroscopy: Fundamentals and Applications*; Ando, D.J., Ed.; John Wiley&Sons: Chichester, UK, 2004; ISBN 978-047-085-427-3.
56. Smith, B.C. The carbonyl group, part V: Carboxylates-coming clean. *Spectroscopy* **2018**, *33*, 20–23.
57. Nyquist, R.A.; Kagel, R.O. Infrared spectra of inorganic compounds. In *Handbook of Infrared and Raman Spectra of Inorganic Compounds and Organic Salts*, 1st ed.; Acad. Press: London, UK, 1971; pp. 1–18, ISBN 978-008-087-852-2.
58. NIST Chemistry WebBook. Available online: http://webbook.nist.gov/chemistry/ (accessed on 2 November 2020).
59. Ocana, M.; Fornés, V.; García Ramos, J.V.; Serna, C.J. Factors affecting the infrared and raman spectra of rutile powders. *J. Solid State Chem.* **1988**, *75*, 364–372. [CrossRef]
60. Savchyn, P.; Karbovnyk, I.; Vistovskyy, V.; Voloshinovskii, A.; Pankratov, V.; Cestelli Guidi, M.; Mirri, C.; Myahkota, O.; Riabtseva, A.; Mitina, N. Vibrational properties of $LaPO_4$ nanoparticles in mid-and far-infrared domain. *J. Appl. Phys.* **2012**, *112*, 124309. [CrossRef]
61. Balasubramanian, C.; Bellucci, S.; Cinque, G.; Marcelli, A.; Guidi, M.C.; Piccinini, M.; Popov, A.; Soldatov, A.; Onorato, P. Characterization of aluminium nitride nanostructures by XANES and FTIR spectroscopies with synchrotron radiation. *J. Phys. Condens. Matter* **2006**, *18*, S2095–S2104. [CrossRef]
62. Bellucci, S.; Popov, A.I.; Balasubramanian, C.; Cinque, G.; Marcelli, A.; Karbovnyk, I.; Savchyn, V.; Krutyak, N. Luminescence, vibrational and XANES studies of AlN nanomaterials. *Radiat. Meas.* **2007**, *42*, 708–711. [CrossRef]

Article

Hydrothermal Synthesis, Crystal Structure, and Spectroscopic Properties of Pure and Eu^{3+}-Doped NaY[SO$_4$]$_2$ · H$_2$O and Its Anhydrate NaY[SO$_4$]$_2$

Constantin Buyer [1], David Enseling [2], Thomas Jüstel [2] and Thomas Schleid [1,*]

[1] Institute for Inorganic Chemistry, University of Stuttgart, D-70569 Stuttgart, Germany; buyer@iac.uni-stuttgart.de

[2] Department of Chemical Engineering, FH Münster University of Applied Sciences, D-48565 Steinfurt, Germany; david.enseling@fh-muenster.de (D.E.); tj@fh-muenster.de (T.J.)

* Correspondence: schleid@iac.uni-stuttgart.de; Tel.: +49-711/685-64240

Citation: Buyer, C.; Enseling, D.; Jüstel, T.; Schleid, T. Hydrothermal Synthesis, Crystal Structure, and Spectroscopic Properties of Pure and Eu^{3+}-Doped NaY[SO$_4$]$_2$ · H$_2$O and Its Anhydrate NaY[SO$_4$]$_2$. *Crystals* **2021**, *11*, 575. https://doi.org/10.3390/cryst11060575

Academic Editors: Aleksej Zarkov, Aivaras Kareiva and Loreta Tamasauskaite-Tamasiuniate

Received: 27 April 2021
Accepted: 17 May 2021
Published: 21 May 2021

Publisher's Note: MDPI stays neutral with regard to jurisdictional claims in published maps and institutional affiliations.

Copyright: © 2021 by the authors. Licensee MDPI, Basel, Switzerland. This article is an open access article distributed under the terms and conditions of the Creative Commons Attribution (CC BY) license (https://creativecommons.org/licenses/by/4.0/).

Abstract: The water-soluble colorless compound NaY[SO$_4$]$_2$ · H$_2$O was synthesized with wet methods in a Teflon autoclave by adding a mixture of Na$_2$[SO$_4$] and Y$_2$[SO$_4$]$_3$ · 8 H$_2$O to a small amount of water and heating it up to 190 °C. By slow cooling, single crystals could be obtained and the trigonal crystal structure of NaY[SO$_4$]$_2$ · H$_2$O was refined based on X-ray diffraction data in space group $P3_221$ ($a = 682.24(5)$ pm, $c = 1270.65(9)$ pm, $Z = 3$). After its thermal decomposition starting at 180 °C, the anhydrate NaY[SO$_4$]$_2$ can be obtained with a monoclinic crystal structure refined from powder X-ray diffraction data in space group $P2_1/m$ ($a = 467.697(5)$ pm, $b = 686.380(6)$ pm, $c = 956.597(9)$ pm, $β = 96.8079(5)$, $Z = 2$). Both compounds display unique Y^{3+}-cation sites with eightfold oxygen coordination (d(Y–O$_s$) = 220–277 pm)) from tetrahedral [SO$_4$]$^{2-}$ anions (d(S–O) = 141–151 pm)) and a ninth oxygen ligand from an H$_2$O molecule (d(Y–O$_w$) = 238 pm) in the hydrate case. In both compounds, the Na$^+$ cations are atoms (d(Na–O$_s$) = 224–290 pm) from six independent [SO$_4$]$^{2-}$ tetrahedra each. Thermogravimetry and temperature-dependent PXRD experiments were performed as well as IR and Raman spectroscopic studies. Eu^{3+}-doped samples were investigated for their photoluminescence properties in both cases. The quantum yield of the red luminescence for the anhydrate NaY[SO$_4$]$_2$:Eu^{3+} was found to be almost 20 times higher than the one of the hydrate NaY[SO$_4$]$_2$ · H$_2$O:Eu^{3+}. The anhydrate NaY[SO$_4$]$_2$:Eu^{3+} exhibits a decay time of about $τ_{1/e} = 2.3$ μm almost independent of the temperature between 100 and 500 K, while the CIE1931 color coordinates at $x = 0.65$ and $y = 0.35$ are very temperature-consistent too. Due to these findings, the anhydrate is suitable as a red emitter in lighting for emissive displays.

Keywords: sodium yttrium oxosulfate; X-ray diffraction; crystal structure; rare-earth metal compounds; luminescence; temperature- and time-dependent photoluminescence

1. Introduction

Eu^{3+}-doped luminescence materials based on complex oxides are very important in application [1,2] and show a red emission with typical $^5D_0 \rightarrow {}^7F_J$ transitions between 610 and 620 nm [3]. They could be prepared on "classic" solid-state routes at high temperatures, as has been done for the examples of Y$_2$[MoO$_4$]$_3$:Eu^{3+} and Y$_2$[MoO$_4$]$_2$[Mo$_2$O$_7$]:Eu^{3+} [4], GdSb$_2$O$_4$Br:Eu^{3+} [5], as well as YNbO$_4$:Eu^{3+} and YTaO$_4$:Eu^{3+} [6]. Another energy-saving synthesis route to get Eu^{3+}-doped luminescence materials without heating uses wet synthesis strategies. For example the Eu^{3+}-doped xenotime-type yttrium oxoarsenate Y[AsO$_4$]:Eu^{3+} [7], the oxophosphate Y[PO$_4$]:Eu^{3+} [8], and the oxocarbonate Y$_2$[CO$_3$]$_3$:Eu^{3+} · n H$_2$O [9] were synthesized following a wet route.

With NaCe[SO$_4$]$_2$ · H$_2$O (trigonal, $P3_121$), Lindgren reported for the first time in 1977 the crystal structure of a sodium rare-earth metal oxosulfate monohydrate yielded from a wet synthesis by adding Ce[OH]$_3$ and Na$_2$[SO$_4$] to aqueous sulfuric acid (H$_2$SO$_4$) and heating it to 230 °C for 7 days [10]. In later works of other groups, the crystal structure of the sodium rare-earth (RE) metal oxosulfate monohydrates NaRE[SO$_4$]$_2$ · H$_2$O for the elements RE = La—Nd and Sm—Dy was solved either in space group $P3_121$ (no. 152) or its enantiomorphic analog $P3_221$ (no. 154). The samples were produced by different syntheses routes [11–18] and an Indian paper from 1989 deals with these double sulfates of trivalent plutonium, as well as the rare-earth metals RE = La—Nd, Sm—Yb, and Y, but does not give some detailed crystallographic information except for the space group derived from powder X-ray diffraction data [19].

By changing the alkali metal sodium to the next bigger one, potassium namely, KLa[SO$_4$]$_2$ · H$_2$O emerges as the only known alkali-metal rare-earth metal oxosulfate monohydrate with the mentioned trigonal structure [20]. For the smaller rare-earth metals (RE = Ce—Nd, Sm—Dy), the potassium-containing oxosulfate monohydrates KRE[SO$_4$]$_2$ · H$_2$O crystallize monoclinically in space group $P2_1/c$ [20–23] in analogy to the isotypic rubidium compounds RbRE[SO$_4$]$_2$ · H$_2$O with RE = Ce, Gd, Ho and Yb [24–26]. For silver instead of an alkali-metal cation also AgRE[SO$_4$]$_2$ · H$_2$O representatives with the crystal structure of NaCe[SO$_4$]$_2$ · H$_2$O were found [27] and by the exchange of the rare-earth metal cation with trivalent bismuth, its oxosulfate monohydrate NaBi[SO$_4$]$_2$ · H$_2$O [28] shows the same trigonal structure as the related rare-earth metal compounds NaRE[SO$_4$]$_2$ · H$_2$O.

In 2006, the photoluminescence spectrum of NaEu[SO$_4$]$_2$ · H$_2$O (excited at λ = 393 nm) [11] and in 2016 analogous spectra of NaTb[SO$_4$]$_2$ · H$_2$O (excited at λ = 320 nm) and NaDy[SO$_4$]$_2$ · H$_2$O (excited at λ = 387 nm) were measured at room temperature [12]. In 2011, Ce^{3+}- and Tb^{3+}-doped samples of NaY[SO$_4$]$_2$ · H$_2$O were the subject of a luminescence investigation [29]. Moreover, in 2015, the sodium rare-earth metal oxosulfate monohydrates NaRE[SO$_4$]$_2$ · H$_2$O with RE = La, Nd, and Gd could be successfully tested as heterogeneous redox catalysts for the selective oxidation of organic sulfides [13]. It is worth mentioning that NaY[SO$_4$]$_2$ · H$_2$O even occurs as a mineral with the name chinleite-(Y) [30], naturally containing all the lanthanoids with roughly the same size as yttrium. The crystallographic data from a structure refinement in space group $P3_221$ have never been deposited at a common database, however.

For the anhydrous sodium rare-earth metal oxosulfates NaRE[SO$_4$]$_2$, their monoclinic crystal structure was solved in space group $P2_1/m$ for RE = Er [31] and Tm [32] and the triclinic one in space group $P\bar{1}$ for RE = La [33] and Nd [31]. For trivalent gold instead of RE^{3+} cations, the monoclinic crystal structure of NaAu[SO$_4$]$_2$ was described in space group $P2_1/n$ [34], but Au^{3+} in square planar oxygen coordination causes marked topological differences. Not so different from the Na$^+$ analogs, for triclinic AgEu[SO$_4$]$_2$ (space group: $P\bar{1}$) with Ag$^+$ in eightfold oxygen coordination, its Eu^{3+} bulk luminescence was also measured very recently [35].

In the following contribution, we report on the preparation of NaY[SO$_4$]$_2$ · H$_2$O via wet synthesis, its trigonal crystal structure, and the red luminescence of Eu^{3+}-doped samples. After thermal decomposition, we obtained its monoclinic anhydrate NaY[SO$_4$]$_2$, which shows an even stronger red luminescence, when Eu^{3+}-doped.

2. Materials and Methods

2.1. Synthesis

Sodium yttrium oxosulfate monohydrate NaY[SO$_4$]$_2$ · H$_2$O was obtained from a wet synthesis by adding 6.6 mmol Na$_2$[SO$_4$] (ChemPur, 99.9%) and 5.5 mmol Y$_2$[SO$_4$]$_3$ · 8 H$_2$O, which means an excess of Na$_2$[SO$_4$], to about 4 ml demineralized water and heated the obtained wet powder to 190 °C in a 25 ml Teflon autoclave overnight, with a yield only limited by the solubility of the monohydrate. Thus, the yield was about $^2/_3$ of the theoretical possible quantity. It could be increased by evaporating the water, but the change of contamination with Y$_2$[SO$_4$]$_3$ · 8 H$_2$O becomes higher then. By slowly cooling the solution down (5 °C per 1 h), single crystals in a size up to 0.3 mm edge length

(Figure 1) of the water-soluble colorless compound NaY[SO$_4$]$_2$ · H$_2$O could be isolated (Equation (1)) and washed with ethanol (Brüggemann, denaturized with petrol ether). The starting material Y$_2$[SO$_4$]$_3$ · 8 H$_2$O was synthesized by evaporating a solution of Y$_2$O$_3$ (ChemPur, 99.9%) in 96% sulfuric acid H$_2$SO$_4$ (Scharr, pure) according to Equation (2).

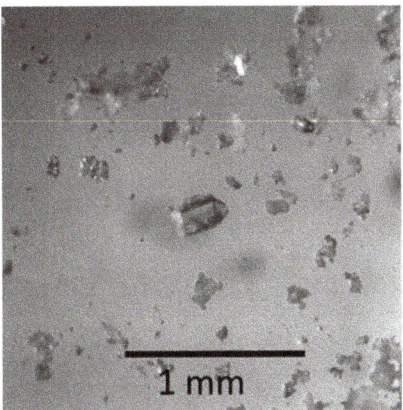

Figure 1. Colorless single crystals of NaY[SO$_4$]$_2$ · H$_2$O.

The anhydrous oxosulfate NaY[SO$_4$]$_2$ can be obtained by heating NaY[SO$_4$]$_2$ · H$_2$O in air at a temperature of 180 °C or higher (Equation (3)). The powder, which was used for the crystal structure refinement, was drained at 550 °C. For the luminescence measurements, a Eu^{3+}-doped sample of NaY[SO$_4$]$_2$ · H$_2$O (0.5% Eu instead of Y) was produced by adding Eu$_2$[SO$_4$]$_3$ · 8 H$_2$O (synthesis analogous to Y$_2$[SO$_4$]$_3$ · 8 H$_2$O with Eu$_2$O$_3$ (ChemPur, 99.9%) instead of Y$_2$O$_3$) to the process, which is described in Equation (1), and NaY[SO$_4$]$_2$:Eu^{3+} has been prepared by draining the doped sample at 550 °C in air.

$$Na_2[SO_4] + Y_2[SO_4]_3 \cdot 8\,H_2O \rightarrow 2\,NaY[SO_4]_2 \cdot H_2O + 7\,H_2O \quad (1)$$

$$RE_2O_3 + 3\,H_2SO_4 + 5\,H_2O \rightarrow RE_2[SO_4]_3 \cdot 8\,H_2O\;(RE = Y\text{ and Eu}) \quad (2)$$

$$NaY[SO_4]_2 \cdot H_2O \rightarrow NaY[SO_4]_2 + H_2O \uparrow \quad (3)$$

2.2. X-ray Experiments and Crystal-Structure Solution

For single-crystal X-ray diffraction experiments, a suitable crystal was selected under a light microscope and fixed inside of a glass capillary with an outer diameter of 0.1 mm and a length of about 15 mm. The crystal was measured with a κ-CCD four-circle X-ray diffractometer (Bruker Nonius, Karlsruhe, Germany) with Mo-Kα radiation (λ = 71.07 pm) at 293 K (room temperature). Crystal-structure solution and refinement for NaY[SO$_4$]$_2$ · H$_2$O (CSD-2016596) in the trigonal space group $P3_221$ were carried out with the program package SHELX-97 [36,37] by Sheldrick, and the program HABITUS by Bärninghausen and Herrendorf was applied [38] for a numerical absorption correction.

For X-ray powder diffraction (PXRD), part of the sample was fixed on a STADI P diffractometer (Stoe & Cie, Darmstadt, Germany) and measured with Cu-Kα radiation (λ = 154.06 pm) in transmission setting. The monohydrate was measured from 2θ = 10–90° for checking phase purity and the anhydrate NaY[SO$_4$]$_2$ (CSD-2072719) was measured from 2θ = 8–110° for solving its crystal structure in the NaEr[SO$_4$]$_2$-type arrangement [31] with the program FULLPROF [39,40]. The measured powder X-ray diffraction pattern of NaY[SO$_4$]$_2$ · H$_2$O can be seen in Figure 2 (top) and the measured PXRD pattern together with the difference plot of the Rietveld refinement for NaY[SO$_4$]$_2$ is shown in Figure 2 (bottom).

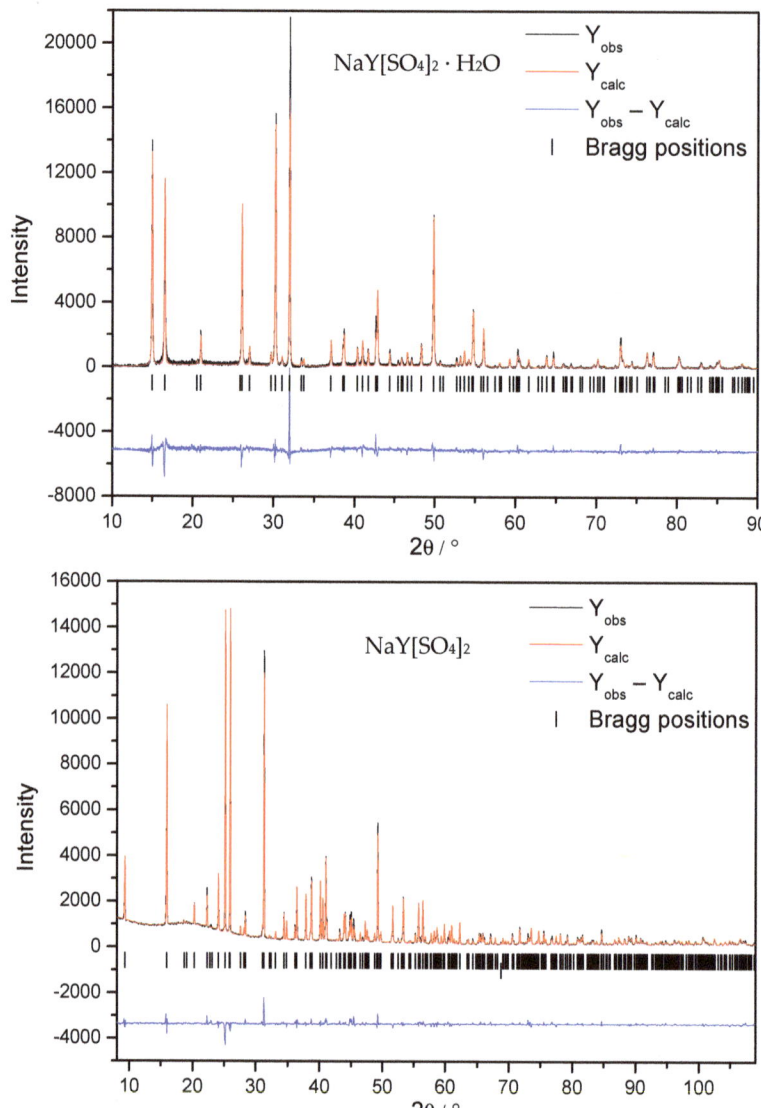

Figure 2. Rietveld refinement based on PXRD data of NaY[SO$_4$]$_2$ · H$_2$O (**top**) for checking its phase purity and NaY[SO$_4$]$_2$ (**bottom**) for crystal-structure determination and refinement.

Temperature-depending powder X-ray diffraction data were measured in the interval 2θ = 10–90° with a RIGAKU SmartLab diffractometer (Neu-Isenburg, Germany) using Cu-Kα radiation (λ = 154.06 pm) in reflection setting from 25 up to 900 °C.

While all the atomic displacement parameters of NaY[SO$_4$]$_2$ · H$_2$O could be refined anisotropically based on single-crystal X-ray diffraction data, the atomic displacement parameters of NaY[SO$_4$]$_2$ were only treated isotropically with Rietveld refinement based on PXRD data.

2.3. Thermal Analysis

Thermal analysis (thermogravimetry) was performed with about 36 mg of a NaY[SO$_4$]$_2$ · H$_2$O samples with a Netzsch device of the type STA-449C (Selb, Germany) in a corundum crucible under argon atmosphere. The sample was heated with 5 K/min from 25 to 1400 °C.

2.4. Luminescence Spectroscopy

Excitation and emission spectra were collected using a fluorescence spectrometer FLS920 (Edinburgh Instruments, Livingston, UK) equipped with a 450 W ozone-free xenon discharge lamp (Osram, München, Germany) and a cryostat "MicrostatN" from Oxford Instruments (Abingdon, UK) as the sample chamber. Additionally, a mirror optic for powder samples was applied. For detection, an R2658P single-photon-counting photomultiplier tube (Hamamatsu, Hamamatsu, Japan) was used. All photoluminescence spectra were recorded with a spectral resolution of 0.5 nm and a dwell time of 0.5 s in 0.5 nm steps.

The photoluminescence decay times were measured on an FLS920 spectrometer (Edinburgh Instruments, Livingston, UK). A Xe μ-flash lamp μF920 was used as an excitation source. For detection, an R2658P single-photon-counting photomultiplier tube (Hamamatsu Photonics, Hamamatsu, Japan) found application.

For the reflection spectra, the investigated samples were placed into an integrating sphere, and FLS920 spectrometer (Edinburgh Instruments, Livingston, UK) equipped with a 450 W Xe lamp, and a cooled (−20 °C) single-photon-counting photomultiplier (Hamamatsu R928) was used. Ba[SO$_4$] was applied as the reflectance standard. The excitation and emission bandwidths were 10.00 and 0.06 nm, respectively. Step width was 0.5 nm and integration time 0.5 s.

Quantum yields were determined according to the method published by Kawamura et al. [41] upon excitation at 395 nm using a 7 nm excitation and 0.5 nm emission slit. The scan steps were 0.5 nm, while the respective emission intensity from 370 to 750 nm was recorded.

The CIE1931 color coordinates and luminous efficacy (*LE*) values were calculated from the temperature-dependent emission spectra of NaY[SO$_4$]$_2$:Eu^{3+} using the Color Calculator 6.75 software from Osram (Osram, München, Germany) [42].

The *LE* value (unit: lm/W) is a parameter describing, how bright the radiation is perceived by an average human observer at a photopic illumination situation. It scales with the photopic human eye sensitivity curve $V(\lambda)$ and can be calculated from the normalized emission spectrum $I(\lambda)$ of the sample as follows [43]:

$$LE(\text{lm/W}) = 683 \, (\text{lm/W}) \cdot \frac{\int_{380nm}^{780nm} I(\lambda)V(\lambda)d\lambda}{\int_{380nm}^{780nm} I(\lambda)d\lambda}$$

2.5. IR and Raman Spectra

Infrared spectra for powder samples of NaY[SO$_4$]$_2$ · H$_2$O and NaY[SO$_4$]$_2$ was measured from 700 to 4000 cm^{-1} with a NICOLET iS5 device from Thermo Scientific (Karlsruhe, Germany). Raman spectroscopy was performed with a DXR SmartRaman spectrometer from Thermo Scientific (Karlsruhe, Germany) with a red laser (λ = 780 nm) and a laser power of 10 mW from 200 to 1800 cm^{-1}.

3. Results and Discussion

3.1. Structure Refinement and Description of NaY[SO$_4$]$_2$ · H$_2$O and NaY[SO$_4$]$_2$

The most relevant crystallographic data of the wet synthesized NaY[SO$_4$]$_2$ · H$_2$O compared to its anhydrate NaY[SO$_4$]$_2$ are shown in Table 1. The given lattice parameters of NaY[SO$_4$]$_2$ · H$_2$O stems from single-crystal data, while its lattice parameters from PXRD experiments amount to a = 682.82(3) pm and c = 1270.77(6) pm (c/a = 1.861).

Table 1. Crystallographic data of NaY[SO$_4$]$_2$ · H$_2$O (**left**) and NaY[SO$_4$]$_2$ (**right**).

Compound	NaY[SO$_4$]$_2$ · H$_2$O	NaY[SO$_4$]$_2$
Crystal system	trigonal	monoclinic
Space group	$P3_221$ (no. 154)	$P2_1/m$ (no. 11)
Lattice parameters,		
a [pm]	682.24(5)	467.697(5)
b [pm]	= a	686.380(6)
c [pm]	1270.65(9)	956.597(9)
β [°]	90	96.8079(5)
Number of formula units, Z	3	2
Unit-cell volume, V_{uc} [nm^3]	0.51219(4)	0.304919(5)
Molar volume, V_m [cm^3 · mol^{-1}]	102.81	91.81
Calculated density, D_x [g · cm^{-3}]	3.132	3.311
Diffraction method	single crystal	powder
Instrument	κ-CCD	Stadi-P (transmission)
Radiation	Mo-Kα, λ = 71.07 pm	Cu-Kα, λ = 154.06 pm
Structure resolution and refinement	SHELX-97	FULLPROF
Range in ±h, ±k, ±l	8, 8, 16	4, 7, 10
Range of 2θ [°]	3–55	8–110
Absorption coefficient, μ [mm^{-1}]	19.25	–
Extinction coefficient, g	0.0174(15)	–
Reflections collected	8159	438
and unique	786	–
R_{int} / R_σ	0.080 / 0.036	–
R_1 / wR_2 for all reflections	0.031 / 0.070	–
Goodness of Fit (GooF)	1.074	–
Residual e$^-$ density (max. / min.)	0.60 and −0.48	–
Flack-x parameter	−0.021(9)	–
R_p	–	4.67
R_{wp}	–	7.52
R_{exp}	–	4.33
χ^2	–	3.02
CSD number	2016596	2072719

Table 2 shows the fractional atomic coordinates with the site symmetry for all atoms and U_{eq} or U_{iso} values of NaY[SO$_4$]$_2$ · H$_2$O and NaY[SO$_4$]$_2$.

While NaY[SO$_4$]$_2$ · H$_2$O crystallizes in the trigonal space group $P3_221$ (no. 154) with a = 682.24(5) pm, and c = 1270.65(9) pm (c/a = 1.862) for Z = 3, NaY[SO$_4$]$_2$ adopts the monoclinic space group $P2_1/m$ (no. 11) with a = 467.697(5) pm, b = 686.380(6) pm, c = 956.597(10) pm, and β = 96.8079(5)° for Z = 2. The b-axes of both compounds differ by only 0.6% and the c-axis of the monohydrate is about $^4/_3$ of the one of the anhydrate. While in the hydrate monolayers of Na$^+$ and Y^{3+} cations take turns along [001], in the anhydrate double layers of each Na$^+$ and Y^{3+} alternate along [001]. Extended unit cells of NaY[SO$_4$]$_2$ · H$_2$O and NaY[SO$_4$]$_2$ can be seen in Figure 3.

In NaY[SO$_4$]$_2$ · H$_2$O, the Y^{3+} cations are coordinated by nine oxygen atoms (eight from oxosulfate anions (d(Y–O) = 237–248 pm) and one from a water molecule (d(Y–O5w) = 238 pm). Only eight oxygen atoms covalently bonded to sulfur in [SO$_4$]$^{2-}$, and units occur as Y^{3+} coordination sphere (d(Y–O) = 220–277 pm) in NaY[SO$_4$]$_2$.

Table 2. Fractional atomic coordinates, site symmetry and U values * of NaY[SO$_4$]$_2 \cdot$ H$_2$O (top) and NaY[SO$_4$]$_2$ (bottom).

Atom	Wyckoff Site	Symmetry	x/a	y/b	z/c	U/pm^2
Na	3b	.2.	0.5299(3)	0	$1/6$	211(5)
Y	3a	.2.	0	0.56341(8)	$1/3$	145(2)
S	6c	1	0.9864(2)	0.5437(2)	0.09243(6)	134(2)
O1	6c	1	0.1273(5)	0.5055(5)	0.0180(2)	217(7)
O2	6c	1	0.8273(5)	0.5829(5)	0.0316(2)	210(7)
O3	6c	1	0.8677(5)	0.3517(5)	0.1655(2)	192(7)
O4	6c	1	0.1249(5)	0.7408(5)	0.1610(2)	196(7)
O5w	3a	.2.	0	0.9123(8)	$1/3$	369(14)
H	6c	1	0.063(11)	0.957(11)	0.042(4)	554(36)
Na	2e	m	0.6289(11)	$1/4$	0.3506(4)	195(12)
Y	2e	m	0.6536(3)	$1/4$	0.82110(12)	167(3)
S1	2e	m	0.1619(7)	$1/4$	0.5875(3)	163(9)
S2	2e	m	0.1407(6)	$1/4$	0.0715(3)	183(9)
O1	2e	m	0.8254(13)	$1/4$	0.0738(6)	114(18)
O2	2e	m	0.2317(13)	$1/4$	0.9259(6)	105(17)
O3	4f	1	0.3075(10)	0.0730(6)	0.6574(4)	177(14)
O4	4f	1	0.2628(10)	0.0699(6)	0.1470(4)	176(13)
O5	2e	m	0.8757(14)	$1/4$	0.6311(6)	119(18)
O6	2e	m	0.1881(13)	$1/4$	0.4401(6)	106(18)

* U values for NaY[SO$_4$]$_2 \cdot$ H$_2$O: $U_{eq} = 1/3\ [U_{33} + 4/3\ (U_{11} + U_{22} - U_{12})]$ [44], but for all atoms of NaY[SO$_4$]$_2$ and H of NaY[SO$_4$]$_2 \cdot$ H$_2$O: U_{iso}.

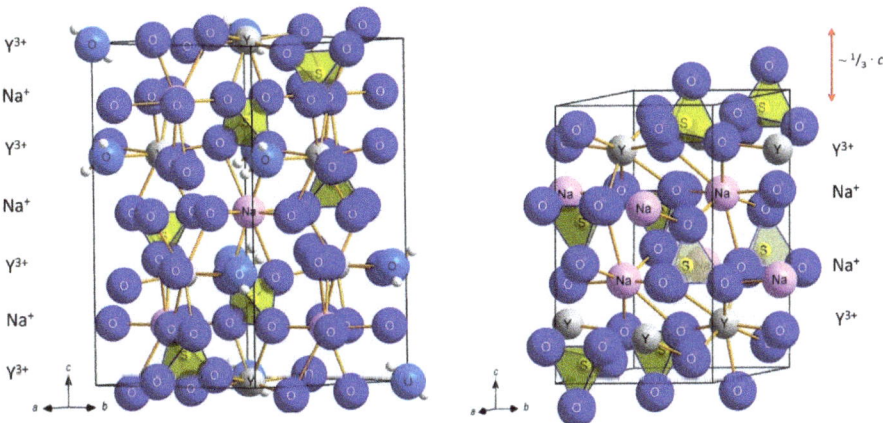

Figure 3. Extended unit cells of NaY[SO$_4$]$_2 \cdot$ H$_2$O (**left**) and NaY[SO$_4$]$_2$ (**right**). While in NaY[SO$_4$]$_2 \cdot$ H$_2$O monolayers of Na$^+$ and Y^{3+} alternate along [001], in NaY[SO$_4$]$_2$ double layers of each Na$^+$ and Y^{3+} do so.

Y^{3+} in NaY[SO$_4$]$_2 \cdot$ H$_2$O resides on the Wyckoff site 3a with C_2 symmetry (Figure 4, left), whereas Y^{3+} in NaY[SO$_4$]$_2$ occupies the 2e position on a mirror plane (Figure 4, right).

In Y$_2$[SO$_4$]$_3 \cdot$ 8 H$_2$O [45], the unique Y^{3+} cations are also surrounded by eight oxygen atoms (four from water molecules and four more from oxosulfate anions) with distances between 230 and 247 pm, while in the anhydrous oxosulfate Y$_2$[SO$_4$]$_3$, Y^{3+} is surrounded octahedrally by only six oxygen atoms from oxosulfate groups with distances between 220 and 224 pm [46]. While Y^{3+} is coordinated by just one oxygen atom per [SO$_4$]$^{2-}$ anion in both Y$_2$[SO$_4$]$_3 \cdot$ 8 H$_2$O [45] and Y$_2$[SO$_4$]$_3$ [46], the same is observed in NaY[SO$_4$]$_2 \cdot$ H$_2$O and NaY[SO$_4$]$_2$, but now with two oxygen atoms of the same oxosulfate unit. In Y$_2$[SO$_4$]$_3$ [46], NaY[SO$_4$]$_2 \cdot$ H$_2$O and NaY[SO$_4$]$_2$ six [SO$_4$]$^{2-}$ anions coordinate the Y^{3+}

cations, while in $Y_2[SO_4]_3 \cdot 8\,H_2O$ [45] there are only four of them. Compounds of the type $ARE[SO_4]_2 \cdot H_2O$ with A = Na crystallize trigonally in space group $P3_221$ (or $P3_121$) [10–18], but monoclinically in space group $P2_1/c$ with A = K for RE = Ce—Nd, Sm—Dy [20–23]. The water molecule in $NaY[SO_4]_2 \cdot H_2O$ is only coordinated to yttrium, whereas in the $KRE[SO_4]_2 \cdot H_2O$ examples [20–23], it further coordinates the alkali-metal cation. The crystal structure of the anhydrous potassium rare-earth metal oxosulfates are described triclinically in space group $P\bar{1}$ for RE = Pr [47] and Nd [48], but monoclinically in space group $P2_1/c$ for RE = Nd [49] and Er [50]. In the triclinic structure, the coordination number of RE^{3+} is eight, while in the monoclinic one, it surprisingly increases to nine. Two oxosulfate anions coordinate with two oxygen atoms each in the monoclinic $KRE[SO_4]_2$ representatives, while in the triclinic cases only one $[SO_4]^{2-}$ group has two contacts to the rare-earth metal cations. The coordination environments of those in the two title compounds are compared to the other alkali-metal rare-earth metal oxosulfates in Figure 5.

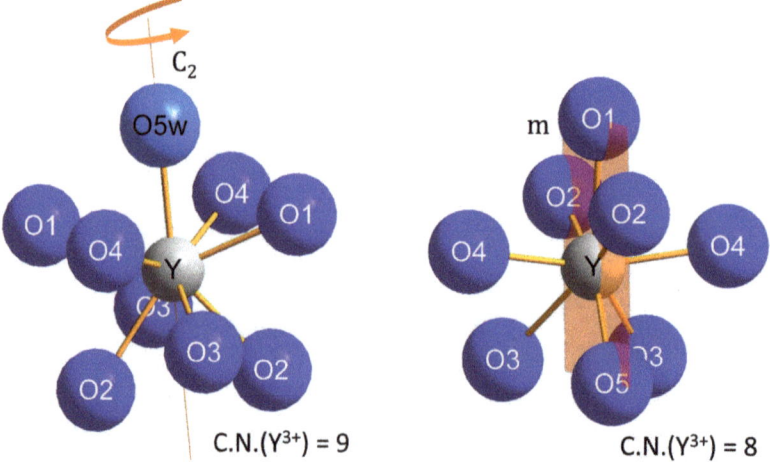

Figure 4. Y^{3+} is coordinated by nine oxygen atoms in $NaY[SO_4]_2 \cdot H_2O$ with C_2 symmetry (**left**) and by eight oxygen atoms in $NaY[SO_4]_2$ residing in a mirror plane (**right**).

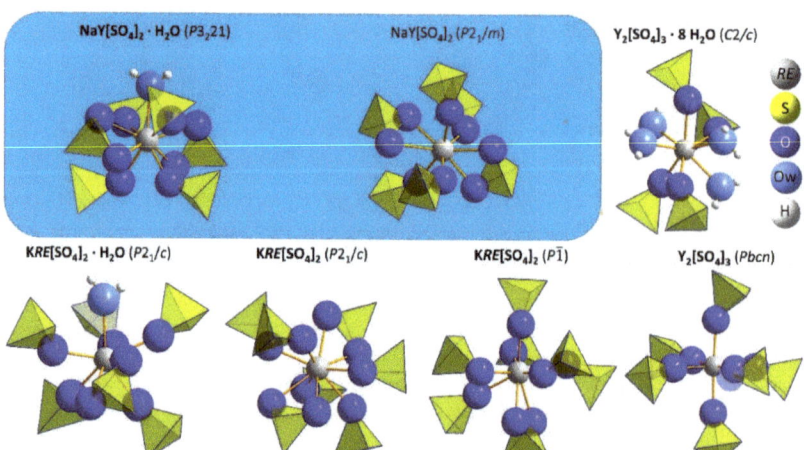

Figure 5. Coordination spheres of the RE^{3+} cations in different rare-earth metal oxosulfates. The blue box contains the title compounds $NaY[SO_4]_2 \cdot H_2O$ and $NaY[SO_4]_2$.

The sodium cations in both title compounds are surrounded by eight oxygen atoms from six different oxosulfate units as a bicapped octahedron. While Na^+ in $NaY[SO_4]_2 \cdot H_2O$ is only connected with $[SO_4]^{2-}$ anions and no water molecules, in the related potassium compound the K^+ cation has contact with six of them and one water molecule [20–23]. The anhydrous potassium rare-earth metal oxosulfates show a coordination sphere around the alkali-metal cation erected by ten oxygen atoms from six oxosulfate anions in case of the triclinic examples [47,48] and seven terminal $[SO_4]^{2-}$ units in the monoclinic cases [49,50]. In the orthorhombic salt $Na_2[SO_4]$, the sodium cations show six oxygen atoms from five oxosulfate groups as next neighbors [51], while in its decahydrate $Na_2[SO_4] \cdot 10\ H_2O$, Na^+ is only surrounded by six water molecules octahedrally [52] (Figure 6).

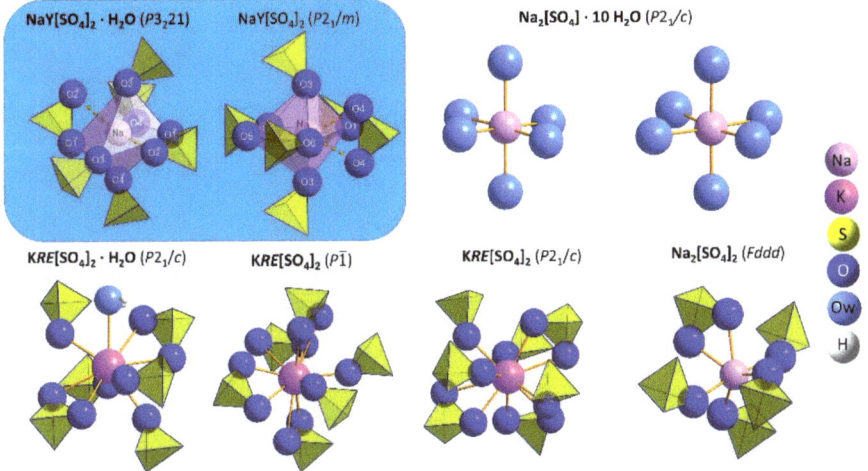

Figure 6. Coordination spheres of the alkali-metal cations (A = Na and K) in different oxosulfates. The blue box contains the title compounds $NaY[SO_4]_2 \cdot H_2O$ and $NaY[SO_4]_2$.

While $NaY[SO_4]_2 \cdot H_2O$ exhibits only one singular crystallographic $[SO_4]^{2-}$ anion, its anhydrate has two different ones of them (Figure 7). All oxygen atoms in $NaY[SO_4]_2 \cdot H_2O$ are surrounded approximately in a plane triangular fashion by Y^{3+}, Na^+, and S^{6+}, while in $NaY[SO_4]_2$ O2 and O6 differ from this scheme since O2 is coordinated by one S^{6+} and two Y^{3+} and O6 by one S^{6+} and two Na^+ cations. Even O5w has one Y^{3+} and two H^+ cations, three neighbors. The triangular environments of the oxygen atoms in $NaY[SO_4]_2 \cdot H_2O$ and $NaY[SO_4]_2$ can be seen in Figure 8. Selected interatomic distances (d/pm) are summarized in Table 3.

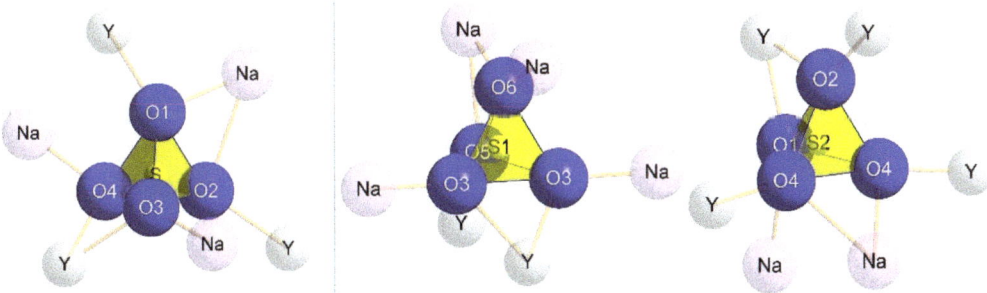

Figure 7. Coordination spheres of the tetrahedral oxosulfate anions $[SO_4]^{2-}$ in $NaY[SO_4]_2 \cdot H_2O$ (**left**) and $NaY[SO_4]_2$ (**right**).

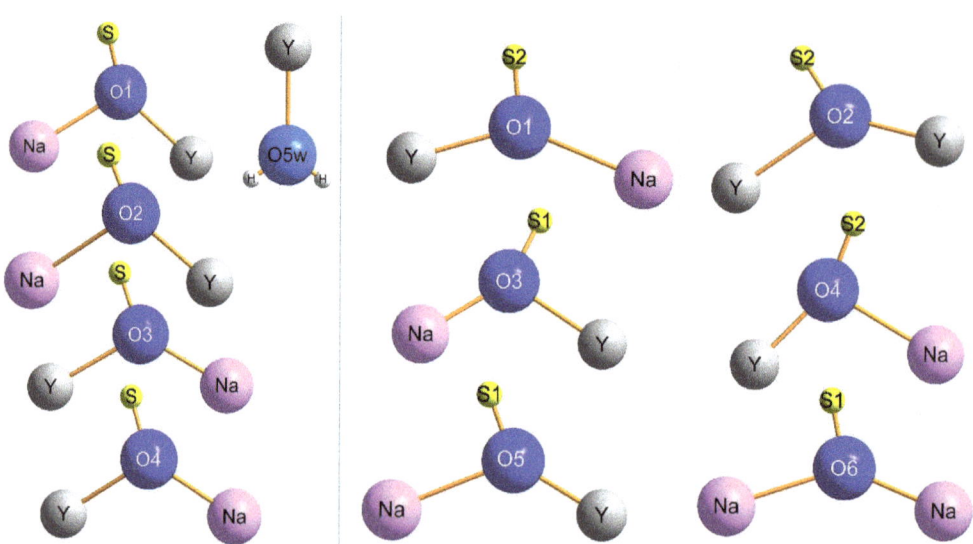

Figure 8. Coordination spheres of the oxygen atoms in NaY[SO$_4$]$_2$ · H$_2$O (**left**) and NaY[SO$_4$]$_2$ (**right**).

Table 3. Selected interatomic distances (d/pm) in the crystal structures of NaY[SO$_4$]$_2$ · H$_2$O (**left**) and NaY[SO$_4$]$_2$ (**right**).

NaY[SO$_4$]$_2$ · H$_2$O			NaY[SO$_4$]$_2$		
d(Y–O2)	(1×)	236.7(3)	d(Y–O5)	(1×)	219.8(7)
d(Y–O2)	(1×)	239.7(4)	d(Y–O4)	(2×)	224.5(4)
d(Y–O5$_W$)	(1×)	238.0(3)	d(Y–O2)	(1×)	231.7(6)
d(Y–O1)	(2×)	239.1(3)	d(Y–O3)	(2×)	243.9(4)
d(Y–O4)	(2×)	224.0(3)	d(Y–O1)	(1×)	245.5(6)
d(Y–O3)	(2×)	247.9(3)	d(Y–O2)	(1×)	277.0(6)
$\bar{d}$(Y–O)	(C.N. = 9)	241.5	$\bar{d}$(Y–O)	(C.N. = 8)	238.8
d(Na–O3)	(2×)	235.4(4)	d(Na–O3)	(2×)	223.9(4)
d(Na–O4)	(2×)	242.5(4)	d(Na–O6)	(1×)	232.4(8)
d(Na–O1)	(2×)	253.6(3)	d(Na–O6)	(1×)	265.4(8)
d(Na–O2)	(2×)	287.9(3)	d(Na–O4)	(2×)	273.2(5)
$\bar{d}$(Na–O)	(C.N. = 8)	254.9	d(Na–O5)	(1×)	279.3(7)
			d(Na–O1)	(1×)	290.4(7)
			$\bar{d}$(Na–O)	(C.N. = 8)	257.7
d(S–O1)	(1×)	146.2(3)			
d(S–O2)	(1×)	146.2(4)	d(S1–O6)	(1×)	141.0(7)
d(S–O3)	(1×)	147.4(3)	d(S1–O5)	(1×)	144.9(8)
d(S–O4)	(1×)	148.0(3)	d(S1–O3)	(2×)	151.0(4)
$\bar{d}$(S–O)	(C.N. = 4)	147.2	$\bar{d}$(S1–O)	(C.N. = 4)	147.4
			d(S2–O1)	(1×)	147.7(7)
			d(S2–O2)	(1×)	150.4(7)
			d(S2–O4)	(2×)	150.9(4)
			$\bar{d}$(S2–O)	(C.N. = 4)	150.0

For confirmation of the Na$^+$ and Y^{3+} sites in NaY[SO$_4$]$_2$ · H$_2$O and NaY[SO$_4$]$_2$, bond-valence calculations were carried out with the parameters used by Brese and O'Keeffe [53]. With calculated charges of 3.06 in NaY[SO$_4$]$_2$ · H$_2$O and 3.19 in NaY[SO$_4$]$_2$ for the Y^{3+} sites next to 1.20 in NaY[SO$_4$]$_2$ · H$_2$O and 1.23 in NaY[SO$_4$]$_2$ for the Na$^+$ sites, their positions can just be confirmed. More details of these calculations can be seen in Table 4. The bond-valence equation for the calculation of the charge given by Brese and O'Keeffe [53] is v_{ij} = $\exp[(R_{ij} - d_{ij})/b]$ with the valence v_{ij}, the universal constant $b = 0.37$ Å, the bond-valence parameter R_{ij}, and the Ångström distance of the considered atoms d_{ij} between the atoms i and j. The sum $\Sigma(v_{ij})$ represents the charge of the regarded ion.

Table 4. Results of the bond-valence calculations for NaY[SO$_4$]$_2$ · H$_2$O and NaY[SO$_4$]$_2$.

NaY[SO$_4$]$_2$ · H$_2$O for Y	O2	O2'	O5	O1	O1'	O4	O4'	O3	O3'	
d(Y–O) [pm]	236.68	236.70	238.03	239.12	239.13	243.98	244.06	247.86	247.93	$\Sigma(v_{ij})$
v_{ij}	0.385	0.385	0.372	0.361	0.361	0.316	0.316	0.285	0.284	3.065
for Na	O3	O3'	O4	O4'	O1	O1'	O2	O2'		
d(Na–O) [pm]	235.35	235.35	242.50	242.50	253.58	253.66	287.92	287.99		$\Sigma(v_{ij})$
v_{ij}	0.224	0.224	0.185	0.185	0.137	0.137	0.054	0.054		1.199
for S	O1	O2	O3	O4		for H		O5w		
d(S–O) [pm]	101.46	101.46	101.47	101.48	$\Sigma(v_{ij})$	d(H–O) [pm]		97.86		
v_{ij}	1.550	1.549	1.500	1.477	6.076	v_{ij}		0.926		
NaY[SO$_4$]$_2$ for Y	O5	O4	O4	O2	O3	O3	O1	O2		
d(Y–O) [pm]	219.77	224.48	224.48	231.68	243.87	243.87	245.53	277.00		$\Sigma(v_{ij})$
v_{ij}	0.609	0.536	0.536	0.441	0.317	0.317	0.303	0.130		3.189
for Na	O3	O3'	O6	O6'	O4	O4'	O5	O1		
d(Na–O) [pm]	223.94	223.94	232.42	265.37	272.18	273.18	279.25	290.39		$\Sigma(v_{ij})$
v_{ij}	0.305	0.305	0.242	0.100	0.083	0.081	0.068	0.051		1.234
for S1	O6	O5	O3 (2×)		for S2		O1	O2	O4 (2×)	
d(S1–O) [pm]	142.97	144.85	150.96	$\Sigma(v_{ij})$	d(S2–O) [pm]		147.74	150.39	150.93	$\Sigma(v_{ij})$
v_{ij}	1.691	1.607	1.362	6.022	v_{ij}		1.486	1.383	1.363	5.597
R_{ij} constant from [53] for		Y	Na	S	H					
distance to O		2.014	1.80	1.624	0.95	Å				

The motifs of mutual adjunction for the atoms in both title compounds NaY[SO$_4$]$_2$ · H$_2$O and NaY[SO$_4$]$_2$ can be seen in Table 5.

Table 5. Motifs of mutual adjunction for NaY[SO$_4$]$_2$ · H$_2$O (top) and NaY[SO$_4$]$_2$ (bottom).

NaY[SO$_4$]$_2$ · H$_2$O	O1	O2	O3	O4	O5w	C.N.
Y	2/1	2/1	2/1	2/1	1/1	9
Na	2/1	2/1	2/1	2/1	0/0	8
S	1/1	1/1	1/1	1/1	0/0	4
H	0/0	0/0	0/0	0/0	1/2	1
C.N.	3	3	3	3	3	
NaY[SO$_4$]$_2$	O1	O2	O3	O4	O5	O6	C.N.
Y	1/1	2/2	2/1	2/1	1/1	0/0	8
Na	1/1	0/0	2/1	2/1	1/1	2/2	8
S1	0/0	0/0	2/1	0/0	1/1	1/1	4
S2	1/1	1/1	0/0	2/1	0/0	0/0	4
C.N.	3	3	3	3	3	3	

We became aware of a competing structure refinement for trigonal NaY[SO$_4$]$_2$ · H$_2$O ($a = 681.91(3)$ pm, $c = 1270.35(11)$ pm, $c/a = 1.863$) in space group $P3_121$ that was already in the progress of publication [54], simultaneous to our activities writing this article. The lower CSD deposition number (ours for NaY[SO$_4$]$_2$ · H$_2$O in space group $P3_221$: 2016596 versus the Chinese competitor one for NaY[SO$_4$]$_2$ · H$_2$O in space group $P3_121$: 2058909) should grant us a priority, despite the almost identical results in both papers from the year 2021.

3.2. Thermal Analysis

A thermogravimetrical curve for the decomposition of NaY[SO$_4$]$_2$ · H$_2$O between 25 and 1400 °C is depicted in Figure 9.

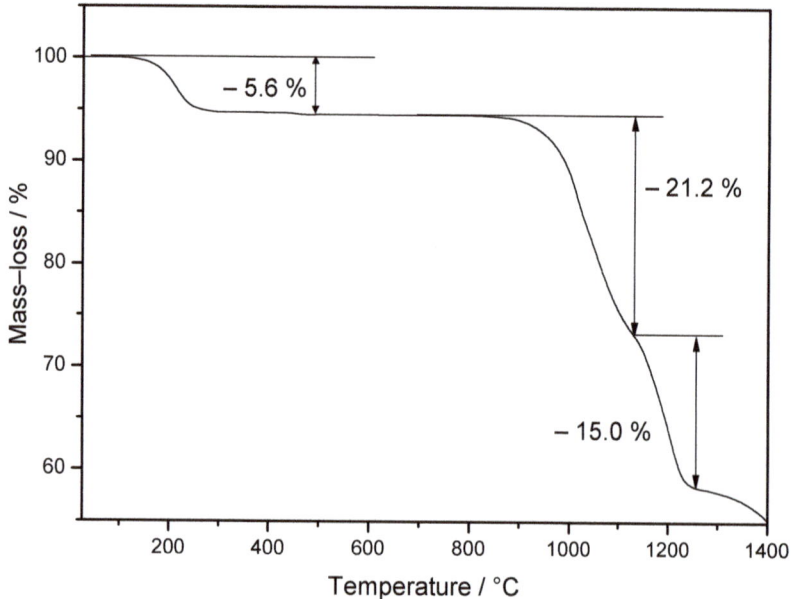

Figure 9. Thermogravimetrical curve of NaY[SO$_4$]$_2$ · H$_2$O between 25 and 1400 °C.

The first mass-loss at about 180 °C with 5.6% represents the release of water and the transformation from NaY[SO$_4$]$_2$ · H$_2$O (100% mass; M = 322.036 g/mol) to NaY[SO$_4$]$_2$ (94.4% mass; M = 304.021 g/mol). The second decomposition leads to a mixture of Y$_2$O$_2$[SO$_4$] [55] with the crystal structure of monoclinic La$_2$O$_2$[SO$_4$] [56] and Eu$_2$O$_2$[SO$_4$] [57] or orthorhombic Nd$_2$O$_2$[SO$_4$] [58] together with Na$_2$[SO$_4$] [51,59], confirmed by powder X-ray diffraction experiments (Figures S1 and S2 in the Supplementary Information). For Y$_2$O$_2$[SO$_4$], there are no known or other good crystal-structure data available, so there are differences in intensity and position, but the final decomposition step leads to a mixture of cubic Y$_2$O$_3$ with a bixbyite-type structure [60] and orthorhombic Na$_2$[SO$_4$] [51,59] (Figure S3). The TG curve (Figure 9) appears to be similar to that of NaRE[SO$_4$]$_2$ · H$_2$O with RE = La, Ce, Nd, and Sm, which have been measured in 1994 by Kolcu and Zümreoğlu-Karan [18]. The dehydration temperature of the lanthanum compound is 297 °C and gets lower with decreasing RE^{3+}-cation radius along with the lanthanoid contraction [61]. With a dehydration temperature of 265 °C for the samarium compound, this trend is further confirmed with our yttrium analog at 180 °C.

Additional to the thermogravimetry, temperature-dependent X-ray diffraction experiments were performed (Figure 10).

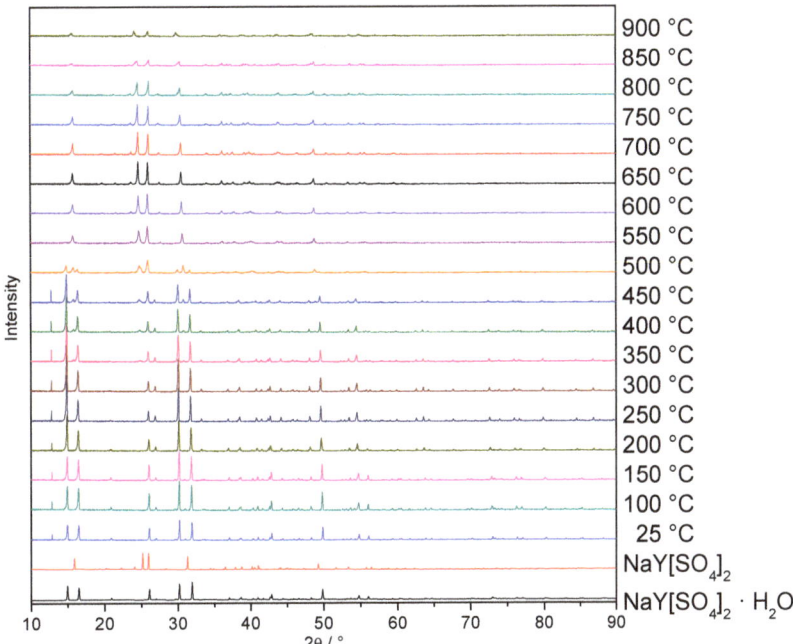

Figure 10. Temperature-dependent PXRD data of NaY[SO$_4$]$_2$ · H$_2$O in the range from 25 to 900 °C measured with Cu-Kα radiation (λ = 154.06 pm) in a reflection setting.

While the TG curve (Figure 9) shows a phase transformation from NaY[SO$_4$]$_2$ · H$_2$O to NaY[SO$_4$]$_2$ at 180 °C, the temperature-depending PXRD indicates the anhydrous compound for the first time at 350 °C. At 550 °C the water-containing compound could not be detected anymore. The XRD intensities became lower again with rising temperatures and suggest a starting decomposition of NaY[SO$_4$]$_2$ to Y$_2$O$_2$[SO$_4$] and Na$_2$[SO$_4$]. The reflection at 12.8° resulted from the X-ray powder-diffractometer setting.

3.3. Luminescence-Spectroscopic Properties

Eu^{3+}-doped samples of NaY[SO$_4$]$_2$ · H$_2$O and NaY[SO$_4$]$_2$ under UV irradiation (λ = 254 nm) can be seen in Figure 11.

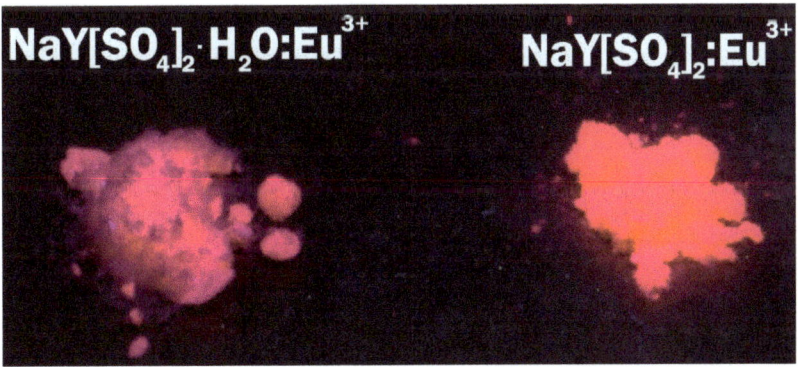

Figure 11. NaY[SO$_4$]$_2$ · H$_2$O:Eu^{3+} (**left**) and NaY[SO$_4$]$_2$:Eu^{3+} (**right**) under UV irradiation (λ_{exc} = 254 nm).

Both compounds display a reflection spectrum, which is in line with plain white powders of good optical quality and high crystallinity, due to the lack of greying or defect bands. The absorption edge of the anhydrous compound at about 270 nm is assigned to the LMCT (ligand-to-metal charge-transfer) absorption band of Eu^{3+}, which is a typical energetic position of the LMCT process of Eu^{3+} in an oxidic environment [62]. The reflectance values at longer wavelengths were close to unity, pointing to a high optical quality of the prepared materials. In both reflection spectra (Figure 12), the typical Eu^{3+} absorption lines originating from the $^7F_0 \rightarrow {}^5L_6$ and $^7F_0 \rightarrow {}^5D_2$ transitions could be observed in the ranges of 395–397 nm and 450–470 nm, respectively [63].

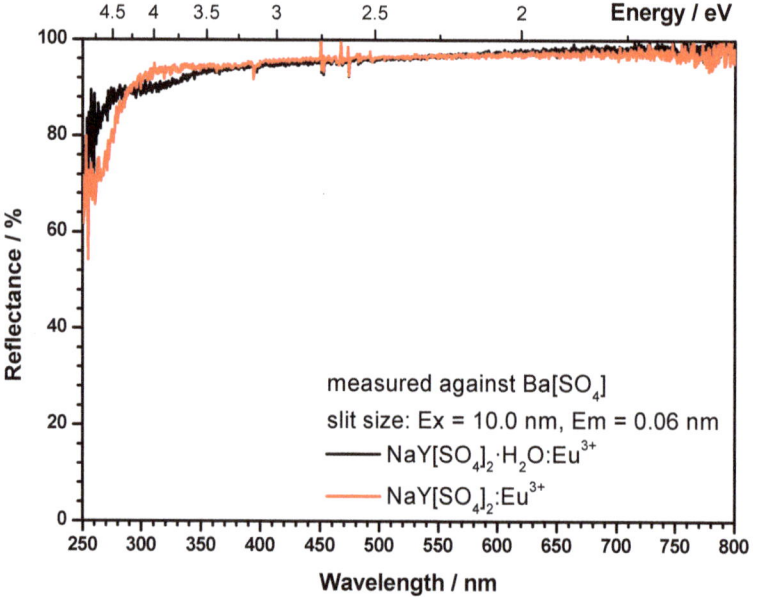

Figure 12. Reflection spectra of $NaY[SO_4]_2 \cdot H_2O:Eu^{3+}$ (black curve) and $NaY[SO_4]_2:Eu^{3+}$ (red curve).

Temperature-dependent excitation spectra of $NaY[SO_4]_2 \cdot H_2O:Eu^{3+}$ and $NaY[SO_4]_2:Eu^{3+}$ reveal the typical intraconfigurational 4f–4f transitions of Eu^{3+} between 280 and 550 nm [63,64] and the position of the LMCT of Eu^{3+} in the anhydrous compound. The LMCT band was located at 270 nm and was in good agreement with the position derived from the reflection spectrum. The excitation spectra of both compounds are plotted in Figure 13.

Noteworthy was the temperature-dependent excitation spectra of $NaY[SO_4]_2 \cdot H_2O:Eu^{3+}$, since a closer look at the UV-A range revealed a distinct change of the pattern of $^7F_0 \rightarrow {}^5L_6$ (390–405 nm) and $^7F_0 \rightarrow {}^5L_8 + {}^5G_J + {}^5L_9 + {}^5L_{10}$ (J = 2–6) (373–387 nm) transitions [65]. The thermal population of the $^7F_1$ level could explain some changes in the excitation line pattern. However, the shift and broadening of the most intense line of the $^7F_0 \rightarrow {}^5L_6$ multiplet at 394 nm from 400 K onwards pointed to a phase transition. This finding could be explained by the loss of water and the transformation of $NaY[SO_4]_2 \cdot H_2O:Eu^{3+}$ to $NaY[SO_4]_2:Eu^{3+}$ in good accordance with the results from thermal gravimetry (Figure 9). Temperature-dependent emission spectra of $NaY[SO_4]_2 \cdot H_2O:Eu^{3+}$ and $NaY[SO_4]_2:Eu^{3+}$ are shown in Figure 14.

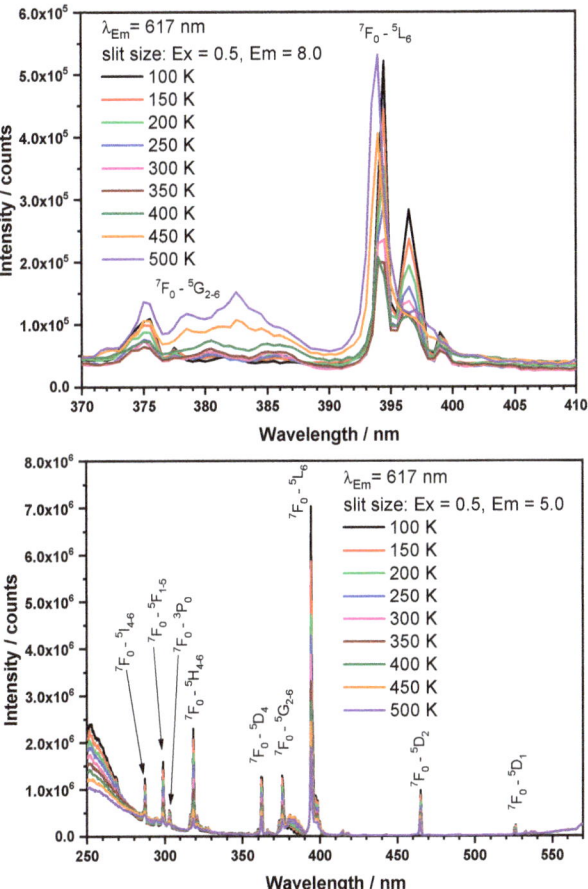

Figure 13. Excitation spectra of NaY[SO$_4$]$_2$ · H$_2$O:Eu^{3+} (**top**) and NaY[SO$_4$]$_2$:Eu^{3+} (**bottom**) as a function of temperature between 100 and 500 K.

The emission spectra of Eu^{3+}-comprising materials consisted of the orange allowed magnetic-dipole (MD) transition $^5D_0 \rightarrow {}^7F_1$, the red parity-forbidden electric-dipole (ED) transition $^5D_0 \rightarrow {}^7F_2$, and further line multiplets in the deep red spectral range around 650 and 695 nm due to the ED transitions $^5D_0 \rightarrow {}^7F_3$ and $^5D_0 \rightarrow {}^7F_4$. For light sources and emissive displays, the emission spectrum should consist mainly of emission lines resulting from the $^5D_0 \rightarrow {}^7F_2$ transitions [66,67]. This means that the Eu^{3+} cation has to occupy a crystallographic site without inversion symmetry (see Figure 4 for symmetry examination). This also induces the deep red emission lines. Fortunately, the $^5D_0 \rightarrow {}^7F_2$ transition is hypersensitive and small deviations of the inversion symmetry strongly enhance the probability of the $^5D_0 \rightarrow {}^7F_2$ transitions. The intensity of the strongly forbidden transition $^5D_0 \rightarrow {}^7F_0$ is known to correlate with the linear terms of the crystal-field parameter and polarizability of the Eu^{3+} cation [67].

However, the emission spectrum of NaY[SO$_4$]$_2$ · H$_2$O:Eu^{3+} upon 395 nm excitation revealed the typical emission line pattern between 580 and 720 nm due to the $^5D_0 \rightarrow {}^7F_J$ (J = 0–4) transitions of Eu^{3+} [62,63,68]. Unfortunately, the signal-to-noise ratio is rather low, which points to a low quantum yield. Indeed, the determination of the quantum efficiency according to Kawamura [41] yielded a value of solely about 1%. Such a low quantum yield can be explained by the presence of crystal water since the high phonon frequency of the O–H vibration of water quenches efficiently the Eu^{3+} luminescence [69].

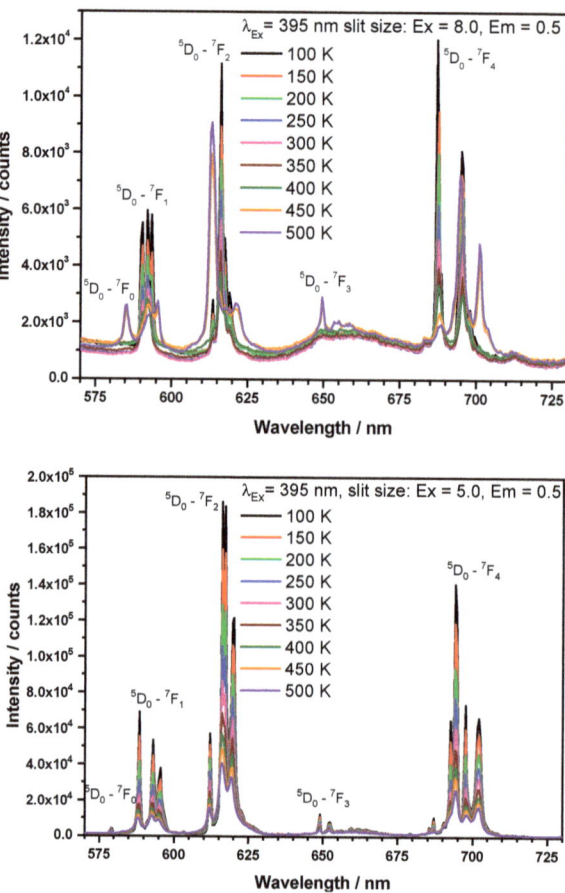

Figure 14. Temperature-dependent emission spectra of $NaY[SO_4]_2 \cdot H_2O:Eu^{3+}$ (**top**) and $NaY[SO_4]_2:Eu^{3+}$ (**bottom**) between 100 and 500 K upon 395 nm excitation.

As already observed for the excitation spectra, the temperature-dependent emission spectra of $NaY[SO_4]_2 \cdot H_2O:Eu^{3+}$ showed a distinct change once the temperature exceeded 400 K, resulting in the increase of intensity and the width of the $^5D_0 \rightarrow {}^7F_1$, $^5D_0 \rightarrow {}^7F_2$, and $^5D_0 \rightarrow {}^7F_4$ transitions, as well as the appearance of the $^5D_0 \rightarrow {}^7F_0$ transition, which was absent at room temperature. This change again points to a phase transition, i.e., the transformation of $NaY[SO_4]_2 \cdot H_2O:Eu^{3+}$ to $NaY[SO_4]_2:Eu^{3+}$, which goes along with an increase of the crystal-field strength causing a larger energetic spread of the Stark components of the above mentioned $^5D_0 \rightarrow {}^7F_J$, transitions. This finding was in good agreement with the decline of the coordination number from 9 to 8 and a shorter average Y–O distance. However, even though the emission spectra of the anhydrous sample obtained after the phase transition resembled that of the as-prepared anhydrous sample, the emission spectra were not completely the same. We assumed that after the phase transition a higher defect density remained, which resulted in line-broadening and a lower signal-to-noise ratio since, without further high-temperature treatment, defects caused by the water removal cannot be healed. In contrast, the as-prepared samples of anhydrous

NaY[SO$_4$]$_2$:Eu^{3+} showed a much higher quantum yield. This value was determined to be almost around 20%, which also explained the much better signal-to-noise ratio of the respective emission spectra as a function of temperature (Figure 14, bottom).

The CIE1931 color coordinates of NaY[SO$_4$]$_2$:Eu^{3+} are at x = 0.65 and y = 0.35, while the temperature impact is rather low, branding the substance as a stable color-consistent material for application in displays or fluorescent light sources [1]. However, the magnification of the color space in Figure 15 demonstrates that the color point shifts slightly to the orange range, which can be caused by the reduction of the asymmetry ratio $^5D_0 \rightarrow {^7F_2}$ /$^5D_0 \rightarrow {^7F_1}$ [63] or by the reduction of the covalency related to the $^5D_0 \rightarrow {^7F_4}$/$^5D_0 \rightarrow {^7F_J}$ ratio [3]. However, both effects are in line with a thermal expansion of the crystals and the Eu^{3+} site causes a decrease of the covalent interaction between Eu^{3+} and oxygen and an increase of the local symmetry.

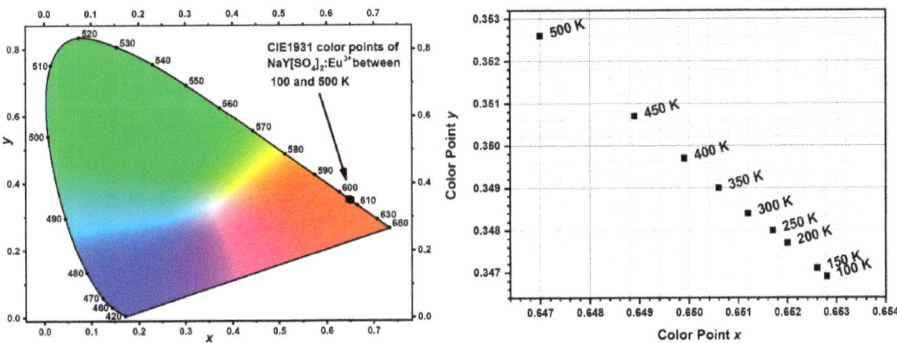

Figure 15. Temperature-dependent CIE1931 color points of the anhydrous NaY[SO$_4$]$_2$:Eu^{3+} between 100 and 500 K upon 395 nm excitation (**left**) and zoom for the magnification of the red area of the color triangle (**right**).

Noteworthy were the intensities of the temperature-dependent emission spectra of NaY[SO$_4$]$_2$ · H$_2$O:Eu^{3+} as depicted in Figure 16. While the intensity decreased between 100 and 300 K due to typical thermal quenching, it increased again between 300 and 500 K. This effect was caused by the phase transition towards the formation of the more efficiently luminescent NaY[SO$_4$]$_2$:Eu^{3+} upon increasing the temperature.

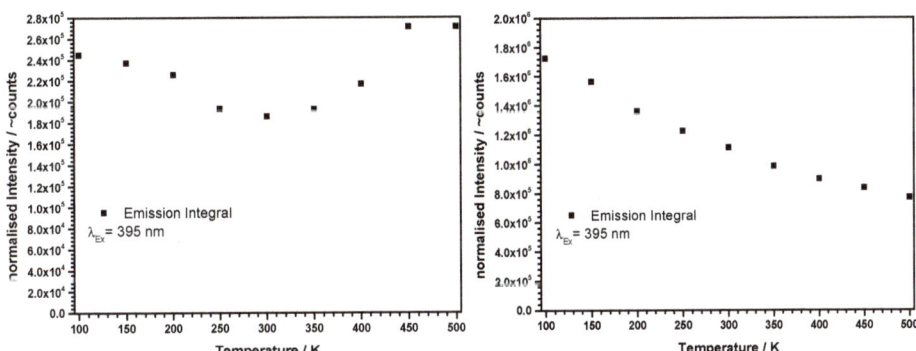

Figure 16. Temperature-dependent emission integrals of NaY[SO$_4$]$_2$ · H$_2$O:Eu^{3+} (**left**) and NaY[SO$_4$]$_2$:Eu^{3+} (**right**) between 100 and 500 K upon 395 nm excitation.

In contrast, the temperature-dependent emission spectra of NaY[SO$_4$]$_2$:Eu^{3+} itself show a typical decrease of the intensity or quantum yield of Eu^{3+} phosphors with increasing temperature [63].

Finally, we investigated the time-dependent luminescence (Figure 17) of the $^5D_0 \rightarrow {}^7F_2$ transition of Eu^{3+} at 617 nm upon 395 nm excitation of $NaY[SO_4]_2 \cdot H_2O:Eu^{3+}$ and $NaY[SO_4]_2:Eu^{3+}$. As discussed before, $NaY[SO_4]_2 \cdot H_2O:Eu^{3+}$ shows a peculiar behavior due to the phase transition between 400 and 500 K, which means that the decay time increases from 550 µs at 100 K to about 930 µs at 500 K. At the same time, the decay curves become bi-exponential, which points to the formation of a novel phase with a prolonged decay time and enhanced internal quantum efficiency. The decay curves of $NaY[SO_4]_2:Eu^{3+}$ between 100 and 500 K were almost perfectly mono-exponential over three orders of magnitude, while the derived decay times remained rather constant, as proven by the just slight decline from 2.35 ms to 2.20 ms. This finding meant that the internal quantum yield stayed quite stable, and thus, thermal quenching of the Eu^{3+} photoluminescence is a minor issue.

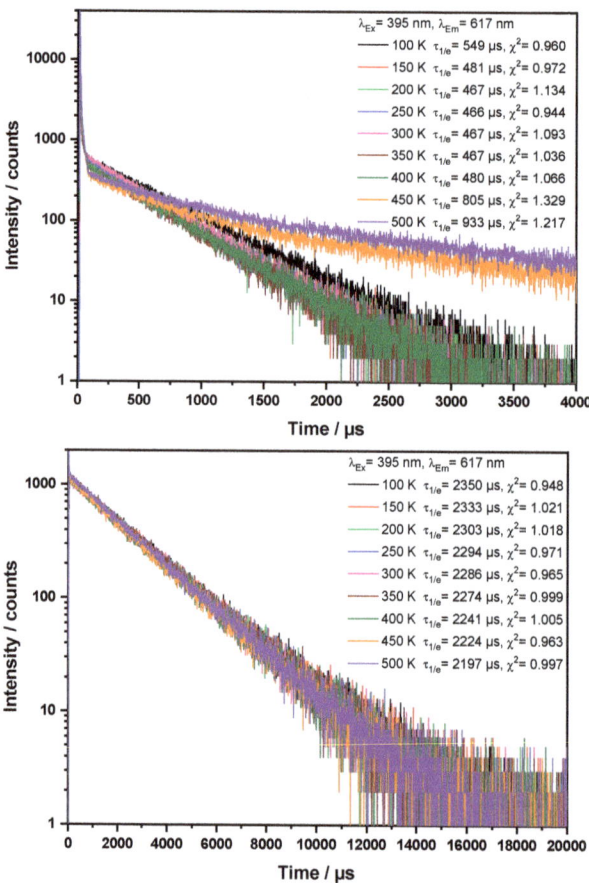

Figure 17. Temperature-dependent decay curves of $NaY[SO_4]_2 \cdot H_2O:Eu^{3+}$ (**top**) and $NaY[SO_4]_2:Eu^{3+}$ (**bottom**) between 100 and 500 K upon 395 nm excitation.

3.4. IR and Raman Studies of $NaY[SO_4]_2 \cdot H_2O$ and $NaY[SO_4]_2$

The Raman and IR spectra of $NaY[SO_4]_2 \cdot H_2O$ and $NaY[SO_4]_2$ are shown together with those of $Y_2[SO_4]_3 \cdot 8 H_2O$ and $Na_2[SO_4]$ in Figure 18 and the values are given in Table 6 compared to the literature data for $Y_2[SO_4]_3$ [69] and $Na_2[SO_4]$ (thenardite) [70]. The vibration at about 2300 cm^{-1} represents CO_2 in the laboratory environment. While the ideal $[SO_4]^{2-}$ anion with T_d symmetry should only have four visible vibration bands (v_{as}

and δ_{as}: IR and Raman active, ν_s and δ_s: only Raman active), in the measured solid-state samples there were more bands measured. This was because of the no longer ideal $[SO_4]^{2-}$ units and their considerable symmetry reduction.

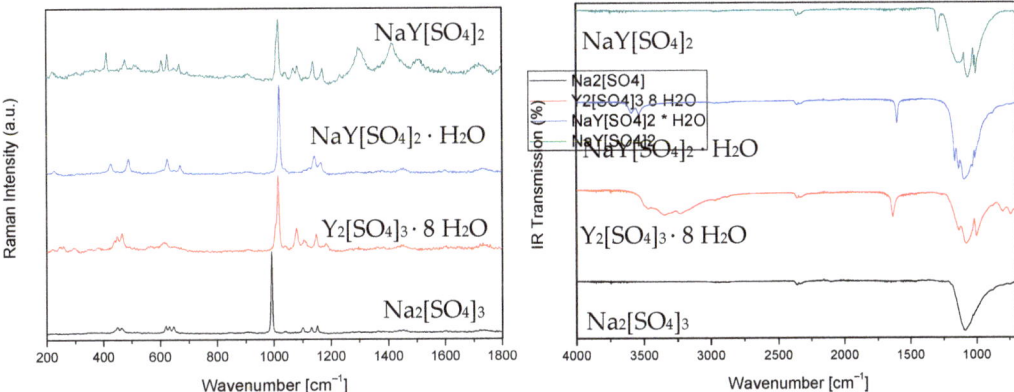

Figure 18. Raman (**left**) and IR spectra (**right**) of $NaY[SO_4]_2$, $NaY[SO_4]_2 \cdot H_2O$, $Y_2[SO_4]_3 \cdot 8\ H_2O$, and $Na_2[SO_4]$ (from top to bottom).

Table 6. Raman (**black**) and IR vibration values (**blue**) of $NaY[SO_4]_2$ and $NaY[SO_4]_2 \cdot H_2O$ as compared to those of $Y_2[SO_4]_3 \cdot 8\ H_2O$, $Y_2[SO_4]_3$ [69], and $Na_2[SO_4]$ (thenardite) [70].

$\tilde{\nu}$ [cm^{-1}]	$NaY[SO_4]_2$	$NaY[SO_4]_2 \cdot H_2O$	$Y_2[SO_4]_3 \cdot 8\ H_2O$	$Y_2[SO_4]_3$ *	$Na_2[SO_4]$ (thenardite) *
$[SO_4]^{2-}$ δ_s	413, 479	429, 492	442, 450, 467	452, 484, 504	451, 466
δ_{as}	608, 629, 669	628, 669	618	609, 654	621, 632, 647
	667, 685	600, 626, 660	638, 652, 688, 743		609, 634, 668
ν_s	1015, 1046	1020	1015	1013	933
	1007, 1010, 1068	1012, 1032, 1092	1002, 1080, 1032		
ν_{as}	1080, 1085, 1140, 1172	1145, 1168	1080, 1088, 1112, 1146	1122, 1145, 1184	1102, 1129, 1152
	1120, 1140, 1289	1135, 1165	1132		1090
H_2O $\nu_{as,s}$		3536, 3592	3229, 3348, 3467		
δ		1606	1640		

* Raman data of $Y_2[SO_4]_3$ and $Na_2[SO_4]$ from literature [69,70], IR data measured in this work.

4. Conclusions

Phase-pure white powder and even colorless single crystals of sodium yttrium oxosulfate monohydrate NaY[SO$_4$]$_2$ · H$_2$O could be synthesized hydrothermally from a mixture of Na$_2$[SO$_4$] and Y$_2$[SO$_4$]$_3$ · 8 H$_2$O in demineralized water. The anhydrate NaY[SO$_4$]$_2$ was obtained by thermal decomposition at temperatures above 180 °C and is stable up to 800 °C. While the trigonal crystal structure of NaY[SO$_4$]$_2$ · H$_2$O was solved from single-crystal X-ray diffraction data in space group $P3_221$, the monoclinic crystal structure of NaY[SO$_4$]$_2$ was refined with *Rietveld* methods from powder X-ray diffraction data in space group $P2_1/m$. The Na$^+$ cations are coordinated by eight oxygen atoms from six tetrahedral [SO$_4$]$^{2-}$ anions in both compounds and the coordination numbers of the Y^{3+} cations in the hydrate amount to nine (eight oxygen atoms from six [SO$_4$]$^{2-}$ units plus one from a water molecule) and eight again in the anhydrate (eight oxygen atoms from six [SO$_4$]$^{2-}$ anions). Both compounds suit as red-emitting luminescent materials, if doped with 0.5 % Eu^{3+}, as shown by luminescence spectroscopy, but the anhydrate NaY[SO$_4$]$_2$:Eu^{3+} exhibits an almost twenty times higher quantum efficiency than the monohydrate NaY[SO$_4$]$_2$ · H$_2$O:Eu^{3+} owing to the water of hydration, which works as a vibrational quencher. The almost perfect monoexponential decay curves of the anhydrate NaY[SO$_4$]$_2$:Eu^{3+} and thus the lack of afterglow also prove the presence of a material with high quality, i.e., a low defect density.

Supplementary Materials: The Supplementary Material contains PXRD data from a sample after thermal treatment from 1000 °C (Figure S1 and S2) and 1400 °C (Figure S3) after the thermogravimetry experiment. They are available online at https://www.mdpi.com/2073-4352/11/6/575/s1.

Author Contributions: C.B. synthesized the pure and the Eu^{3+}-doped compounds, which are described here, and measured their IR and Raman spectra. C.B. and T.S. solved the crystal structures of both compounds. T.S. contributed the reagents, the materials, the scientific equipment, and the infrastructure for synthesis, IR, and Raman spectroscopy. D.E. and T.J. measured and interpreted the Eu^{3+} luminescence of both doped compounds. The paper was written by all authors. All authors have read and agreed to the published version of the manuscript.

Funding: This research was funded by the State of Baden-Württemberg (Stuttgart).

Data Availability Statement: Crystallographic data are available in the ICSD-Database with the CSD-numbers 2016596 for NaY[SO$_4$]$_2$ · H$_2$O and 2072719 for NaY[SO$_4$]$_2$.

Acknowledgments: We thank Falk Lissner for the single-crystal X-ray diffraction measurements, Patrik Djendjur for the thermal analyses, and Jean-Louis Hoslauer for the temperature-dependent PXRD experiments.

Conflicts of Interest: The authors declare no conflict of interest.

References

1. Jüstel, T.; Nikol, H.; Ronda, C. New Developments in the Field of Luminescent Materials for Lighting and Displays. *Angew. Chem. Int. Ed.* **1998**, *37*, 3084–3103. [CrossRef]
2. Feldmann, C.; Jüstel, T.; Ronda, C.R.; Schmidt, P.J. Inorganic Luminescent Materials: 100 Years of Research and Application. *Adv. Funct. Mater.* **2003**, *13*, 511–516. [CrossRef]
3. Skaudzius, R.; Katelnikovas, A.; Enseling, D.; Kareiva, A.; Jüstel, T. Dependence of the $^5D_0 \rightarrow {}^7F_4$ transitions of Eu^{3+} on the local environment in phosphates and garnets. *J. Lumin.* **2014**, *147*, 290–294. [CrossRef]
4. Laufer, S.; Strobel, S.; Schleid, T.; Cybinska, J.; Mudring, A.-V.; Hartenbach, I. Yttrium(III) Oxomolybdates(VI) as Potential Host Materials for Luminescence Applications: An Investigation of Eu^{3+}-Doped Y$_2$[MoO$_4$]$_3$ and Y$_2$[MoO$_4$]$_2$[Mo$_2$O$_7$]. *New J. Chem.* **2013**, *37*, 1919–1926. [CrossRef]
5. Goerigk, F.C.; Paterlini, V.; Dorn, K.V.; Mudring, A.-V.; Schleid, T. Synthesis and Crystal Structure of the Short *Ln*Sb$_2$O$_4$Br Series (*Ln* = Eu–Tb) and Luminescence Properties of Eu^{3+}-Doped Samples. *Crystals* **2020**, *10*, 1089. [CrossRef]
6. Popovic, E.J.; Imre-Lucaci, F.; Muresan, L.; Stefan, M.; Bica, E.; Grecu, R.; Indrea, E. Spectral investigations on niobium and rare earth activated yttrium tantalate powders. *J. Optoelectron. Adv. Mater.* **2008**, *10*, 2334–2337.
7. Ledderboge, F.; Nowak, J.; Massonne, H.-J.; Förg, K.; Höppe, H.A.; Schleid, T. High-Pressure Investigations of Yttrium(III) Oxoarsenate(V): Crystal Structure and Luminescence Properties of Eu^{3+}-Doped Scheelite-Type Y[AsO$_4$] from Xenotime-Type Precursors. *J. Solid State Chem.* **2018**, *263*, 65–71. [CrossRef]

8. Cybińska, J. Temperature dependent morphology variation of red emitting microcrystalline YPO$_4$:Eu^{3+} fabricated by hydrothermal method. *Opt. Mater.* **2017**, *65*, 88–94. [CrossRef]
9. Chang, H.-Y.; Chen, F.-S.; Lu, C.-H. Preparation and luminescence characterization of new carbonate (Y$_2$(CO$_3$)$_3$ · n H$_2$O:Eu^{3+}) phosphors via the hydrothermal route. *J. Alloys Compd.* **2011**, *509*, 10014–10019. [CrossRef]
10. Lindgren, O. The Crystal Structure of Sodium Cerium(III) Sulfate Hydrate, NaCe(SO$_4$)$_2$ · H$_2$O. *Acta Chem. Scand.* **1977**, *31*, 591–594. [CrossRef]
11. Wu, C.-D.; Liu, Z.-Y. Hydrothermal synthesis of a luminescent europium(III) sulfate with three-dimensional chiral framework structure. *J. Solid State Chem.* **2006**, *179*, 3500–3504. [CrossRef]
12. Zhai, B.; Li, Z.; Zhang, C.; Zhang, F.; Zhang, X.; Zhang, F.; Cao, G.; Li, S.; Yang, X. Three rare Ln–Na heterometallic 3D polymers based on sulfate anion: Syntheses, structures, and luminescence properties. *Inorg. Chem. Commun.* **2016**, *63*, 16–19. [CrossRef]
13. Perles, J.; Fortes-Revilla, C.; Gutiérrez-Puebla, E.; Iglesias, M.; Monge, M.Á.; Ruiz-Valero, C.; Snejko, N. Synthesis, Structure, and Catalytic Properties of Rare-Earth Ternary Sulfates. *Chem. Mater.* **2005**, *17*, 2701–2706. [CrossRef]
14. Paul, A.K.; Kanagaraj, R. Synthesis, Characterization, and Crystal Structure Analysis of New Mixed Metal Sulfate NaPr(SO$_4$)$_2$(H$_2$O). *J. Struct. Chem.* **2019**, *60*, 477–484. [CrossRef]
15. Blackburn, A.C.; Gerkin, R.E. Sodium lanthanum(III) sulfate monohydrate, NaLa(SO$_4$)$_2$ · H$_2$O. *Acta Crystallogr.* **1994**, *50*, 835–838. [CrossRef]
16. Blackburn, A.C.; Gerkin, R.E. Redetermination of sodium cerium(III) sulfate monohydrate, NaCe(SO$_4$)$_2$ · H$_2$O. *Acta Crystallogr.* **1995**, *51*, 2215–2218. [CrossRef] [PubMed]
17. Kolcu, Ö.; Zümreoğlu-Karan, B. Nonisothermal Dehydration Kinetics of Sodium-Light Lanthanoid Double Sulfate Monohydrates. *Thermochim. Acta* **1997**, *296*, 135–139. [CrossRef]
18. Kolcu, Ö.; Zümreoğlu-Karan, B. Thermal Properties of Sodium-Light-Lanthanoid Double Sulfate Monohydrates. *Thermochim. Acta* **1994**, *240*, 185–198. [CrossRef]
19. Iyer, P.N.; Natarajan, P.R. Double Sulphates of Plutonium(III) and lanthanides with sodium. *J. Less Common Met.* **1989**, *146*, 161–166. [CrossRef]
20. Kazmierczak, K.; Höppe, H.A. Syntheses, crystal structures and vibrational spectra of KLn(SO$_4$)$_2$ · H$_2$O (Ln = La, Nd, Sm, Eu, Gd, Dy). *J. Solid State Chem.* **2010**, *183*, 2087–2094. [CrossRef]
21. Jemmali, M.; Walha, S.; Ben Hassen, R.; Vaclac, P. Potassium cerium(III) bis(sulfate) monohydrate, KCe(SO$_4$)$_2$ · H$_2$O. *Acta Crystallogr.* **2005**, *61*, i73–i75. [CrossRef]
22. Iskhakova, L.D.; Sarukhanyan, N.L.; Shchegoleva, T.M.; Trunov, V.K. The crystal structure of KPr(SO$_4$)$_2$ · H$_2$O. *Kristallografiya* **1985**, *30*, 474–479.
23. Lyutin, V.I.; Safyanov, Y.N.; Kuzmin, E.A.; Ilyukhin, V.V.; Belov, N.V. Rhomb evaluation of the crystal structure of K-Tb double sulfate. *Kristallografiya* **1974**, *19*, 376–378.
24. Robinson, P.D.; Jasty, S. Rubidium Cerium Sulfate Monohydrate, RbCe(SO$_4$)$_2$ · H$_2$O. *Acta Crystallogr.* **1998**, *C54*, IUC9800021. [CrossRef]
25. Prokofev, M.V. Crystal structure of the double sulfate of rubidium and holmium. *Kristallografiya* **1981**, *26*, 598–600.
26. Sarukhanyan, N.L.; Iskhakova, L.D.; Trunov, V.K.; Ilyukhin, V.V. Crystal structure of RbLn(SO$_4$)$_2$(H$_2$O) (Ln = Gd, Ho, Yb). *Koord. Khim.* **1984**, *10*, 981–987.
27. Audebrand, N.; Auffrédic, J.-P.; Louër, D. Crystal structure of silver cerium sulfate hydrate, AgCe(SO$_4$)$_2$ · H$_2$O. *Z. Kristallogr.* **1998**, *213*, 481. [CrossRef]
28. Cheng, S.; Wu, Y.; Mei, D.; Wen, S.; Doert, T.H. Synthesis, Crystal Structures, Spectroscopic Characterization, and Thermal Analyses of the New Bismuth Sulfates NaBi(SO$_4$)$_2$ · H$_2$O and ABi(SO$_4$)$_2$ (A = K, Rb, Cs). *Z. Anorg. Allg. Chem.* **2020**, *646*, 1688–1695. [CrossRef]
29. Song, Y.; Zou, H.; Sheng, Y.; Zheng, K.; You, H. 3D Hierarchical Architectures of Sodium Lanthanide Sulfates: Hydrothermal Synthesis, Formation Mechanisms, and Luminescence Properties. *J. Phys. Chem. C* **2011**, *115*, 19463–19469. [CrossRef]
30. Kampf, A.R.; Nash, B.P.; Marty, J. Chinleite-(Y), NaY(SO$_4$)$_2$ · H$_2$O, a new rare-earth sulfate mineral structurally related to bassanite. *Mineral. Mag.* **2017**, *81*, 909–916. [CrossRef]
31. Sirotinkine, S.P.; Tchijov, S.M.; Pokrovskii, A.N.; Kovba, L.M. Structure cristalline de sulfates doubles de sodium et de terres rares. *J. Less Common Met.* **1978**, *58*, 101–105. [CrossRef]
32. Chizhov, S.M.; Pokrovskii, A.N.; Kovba, L.M. The crystal structure of alpha-NaTm(SO$_4$)$_2$. *Kristallografiya* **1982**, *27*, 997–998.
33. Chizhov, S.M.; Pokrovskii, A.N.; Kovba, L.M. The crystal structure of NaLa(SO$_4$)$_2$. *Kristallografiya* **1981**, *26*, 834–836.
34. Wickleder, M.S.; Büchner, O. The Gold Sulfates MAu(SO$_4$)$_2$ (M = Na, K, Rb). *Z. Naturforsch.* **2001**, *56*, 1340–1343. [CrossRef]
35. Denisenko, Y.G.; Atuchin, V.V.; Molokeev, M.S.; Aleksandrovsky, A.S.; Krylov, A.S.; Oreshonkov, A.S.; Volkova, S.S.; Andreev, O.V. Structure, Thermal Stability, and Spectroscopic Properties of Triclinic Double Sulfate AgEu(SO$_4$)$_2$ with Isolated SO$_4$ Groups. *Inorg. Chem.* **2018**, *57*, 13279–13288. [CrossRef]
36. Sheldrick, G.M. A short history of SHELX. *Acta Crystallogr.* **2008**, *64*, 112–122. [CrossRef] [PubMed]
37. Sheldrick, G.M. *Program. Suite for the Solution and Refinement of Crystal Structures*; Univeristy of Göttingen: Göttingen, Germany, 1997.
38. Bärnighausen, W.; Herrendorf, H. *Habitus*; Program for the Optimization of the Crystal Shape for Numerical Absorption Correction in X-SHAPE; Karlsruhe: Gießen, Germany, 1993.

39. Rodríguez-Carvajal, J. Recent advances in magnetic structure determination by neutron powder diffraction. *Phys. B Condens. Matter* **1993**, *192*, 55–69. [CrossRef]
40. Roisnel, T.; Rodríguez-Carvajal, J. WinPLOTR: A Windows tool for powder diffraction analysis. In *Materials Science Forum. Proceedings of the European Powder Diffraction Conference*; CiteSeerX: State College, PA, USA, 2001.
41. Kawamura, Y.; Sasabe, H.; Adachi, C. Simple Accurate System for Measuring Absolute Photoluminescence Quantum Efficiency in Organic Solid-State Thin Films. *Jpn. J. Appl. Phys.* **2004**, *43*, 7729–7730. [CrossRef]
42. Downloaded at 10 April 2021. Available online: https://www.osram.us/cb/tools-and-resources/applications/led-colorcalculator/index.jsp (accessed on 27 April 2021).
43. Smet, P.F.; Parmentier, A.B.; Poelman, D. Selecting Conversion Phosphors for White Light-Emitting Diodes. *J. Electrochem. Soc.* **2011**, *158*, 37. [CrossRef]
44. Fischer, R.X.; Tillmanns, E. The equivalent isotropic displacement factor. *Acta Crystallogr.* **1988**, *44*, 775–776. [CrossRef]
45. Held, P.; Wickleder, M.S. Yttrium(III) sulfate octahydrate. *Acta Crystallogr.* **2003**, *59*, i98–i100. [CrossRef]
46. Wickleder, M.S. Wasserfreie Sulfate der Selten-Erd-Elemente: Synthese und Kristallstruktur von $Y_2(SO_4)_3$ und $Sc_2(SO_4)_3$. *Z. Anorg. Allg. Chem.* **2000**, *626*, 1468–1472. [CrossRef]
47. Degtyarev, P.A.; Pokrovskii, A.N.; Kovba, L.M. Crystal structure of the anhydrous double sulfate $KPr(SO_4)_2$. *Kristallografiya* **1978**, *23*, 840–843.
48. Degtyarev, P.A.; Korytnaya, F.M.; Pokrovskii, A.N.; Kovba, L.M. Crystal structure of the anhydrous double sulfate of potassium and neodymium $KNd(SO_4)_2$. *Vestn. Mosk. Univ. Seriya 2 Khimiya* **1997**, *16*, 705–708.
49. Iskhakova, L.D.; Gasanov, Y.M.; Trunov, V.K. Crystal structure of the monoclinic modification of $KNd(SO_4)_2$. *J. Struct. Chem.* **1988**, *29*, 242–246. [CrossRef]
50. Sarukhanyan, N.L.; Iskhakova, L.D.; Trunov, V.K. The crystal structure of $KEr(SO_4)_2$. *Kristallografiya* **1985**, *30*, 274–278.
51. Zachariasen, W.H.; Ziegler, G.E. The crystal structure of anhydrous sodium sulfate Na_2SO_4. *Z. Kristallogr.* **1932**, *81*, 92–101. [CrossRef]
52. Ruben, H.W.; Templeton, D.H.; Rosenstein, R.D.; Olovsson, I. Crystal Structure and Entropy of Sodium Sulfate Decahydrate. *J. Am. Chem. Soc.* **1961**, *83*, 820–824. [CrossRef]
53. Brese, N.E.; O'Keeffe, M. Bond-valence parameters for solids. *Acta Crystallogr.* **1991**, *47*, 192. [CrossRef]
54. Wu, C.; Lin, L.; Wu, T.; Huang, Z.; Zhang, C. Deep-ultraviolet transparent alkali metal-rare earth metal sulfate $NaY(SO_4)_2 \cdot H_2O$ as a nonlinear optical crystal: Synthesis and characterization. *CrystEngComm* **2021**, *23*, 2945–2951. [CrossRef]
55. Kijima, T.; Shinbori, T.; Sekita, M.; Uota, M.; Sakai, G. Abnormally enhanced Eu^{3+} emission in $Y_2O_2SO_4$:Eu^{3+} inherited from their precursory dodecylsulfate-templated concentric-layered nanostructure. *J. Lumin.* **2008**, *128*, 311–316. [CrossRef]
56. Zhukov, S.; Yatsenko, A.; Chernyshev, V.; Trunov, V.; Tserkovnaya, E.; Antson, O.; Hölsä, J.; Baulés, P.; Schenk, H. Structural study of lanthanum oxysulfate $(LaO)_2SO_4$. *Mater. Res. Bull.* **1997**, *32*, 43–50. [CrossRef]
57. Hartenbach, I.; Schleid, T. Serendipitous Formation of Single-Crystalline $Eu_2O_2[SO_4]$. *Z. Anorg. Allg. Chem.* **2002**, *628*, 2171. [CrossRef]
58. Golovnev, N.N.; Molokeev, M.S.; Vereshchagin, S.N.; Atuchin, V.V. Synthesis and thermal transformation of a neodymium(III) complex $[Nd(HTBA)_2(C_2H_3O_2)(H_2O)_2] \cdot 2 H_2O$ to non-centrosymmetric oxosulfate $Nd_2O_2SO_4$. *J. Coord. Chem.* **2015**, *68*, 1865–1877. [CrossRef]
59. Niggli, A. Die Raumgruppe von Na_2CrO_4. *Acta Crystallogr.* **1954**, *7*, 776. [CrossRef]
60. Brauer, G.; Gradinger, H. Über heterotype Mischphasen bei Seltenerdoxyden. *Z. Anorg. Allg. Chem.* **1954**, *276*, 209–226. [CrossRef]
61. Shannon, R.D. Revised effective ionic radii and systematic studies of interatomic distances in halides and chalcogenides. *Acta Crystallogr.* **1976**, *32*, 751. [CrossRef]
62. Blasse, B.; Grabmaier, C. *Luminescent Materials*; Springer: Berlin/Heidelberg, Germany; New York, NY, USA, 1994; ISBN 3-540-58019-0.
63. Binnemans, K. Interpretation of europium(III) spectra. *Coord. Chem. Rev.* **2015**, *295*, 1–45. [CrossRef]
64. Dorenbos, P. The Eu^{3+} charge transfer energy and the relation with the band gap of compounds. *J. Lumin.* **2005**, *111*, 89–104. [CrossRef]
65. Baur, F.; Jüstel, T. New Red-Emitting Phosphor $La_2Zr_3(MoO_4)_9$:Eu^{3+} and the Influence of Host Absorption on its Luminescence Efficiency. *Aust. J. Chem.* **2015**, *68*, 1727–1734. [CrossRef]
66. Baur, F.; Jüstel, T. Eu^{3+} activated molybdates—Structure property relations. *Opt. Mater. X* **2019**, *1*, 100015. [CrossRef]
67. Blasse, G. Reminiscencies of a quenched luminescence investigatory. *J. Lumin.* **2002**, *100*, 65–67. [CrossRef]
68. Frech, R.; Cole, R.; Dharmasena, G. Raman Spectroscopic Studies of $Y_2(SO_4)_3$ Substitution in $LiNaSO_4$ and $LiKSO_4$. *J. Solid State Chem.* **1993**, *105*, 151–160. [CrossRef]
69. Supkowski, R.M.; Horrocks, W.D.W. On the determination of the number of water molecules, q, coordinated to europium(III) ions in solution from luminescence decay lifetimes. *Inorg. Chim. Acta* **2002**, *340*, 44–48. [CrossRef]
70. Prieto-Taboada, N.; Fdez-Ortiz de Vallejuelo, S.; Veneranda, M.; Lama, E.; Castro, K.; Arana, G.; Larrañaga, A.; Madariaga, J.M. The Raman spectra of the Na_2SO_4–K_2SO_4 system: Applicability to soluble salts studies in built heritage. *J. Raman Spectrosc.* **2019**, *50*, 175–183. [CrossRef]

Article

Characterization of GAGG Doped with Extremely Low Levels of Chromium and Exhibiting Exceptional Intensity of Emission in NIR Region

Greta Inkrataite [1,*], Gerardas Laurinavicius [1], David Enseling [2], Aleksej Zarkov [1], Thomas Jüstel [2] and Ramunas Skaudzius [1]

1. Faculty of Chemistry and Geosciences, Institute of Chemistry, Vilnius University, Naugarduko 24, LT-03225 Vilnius, Lithuania; gerardas.laurinavicius@chgf.stud.vu.lt (G.L.); aleksej.zarkov@chf.vu.lt (A.Z.); ramunas.skaudzius@chgf.vu.lt (R.S.)
2. Department of Chemical Engineering, Münster University of Applied Sciences, Stegerwaldstrasse 39, D-48565 Steinfurt, Germany; david.enseling@fh-muenster.de (D.E.); tj@fh-muenster.de (T.J.)
* Correspondence: greta.inkrataite@chgf.vu.lt; Tel.: +370-5-219-3105

Abstract: Cerium and chromium co-doped gadolinium aluminum gallium garnets were prepared using sol-gel technique. These compounds potentially can be applied for NIR-LED construction, horticulture and theranostics. Additionally, magnesium and calcium ions were also incorporated into the structure. X-ray diffraction data analysis confirmed the all-cubic symmetry with an Ia-3d space group, which is appropriate for garnet-type materials. From the characterization of the luminescence properties, it was confirmed that both chromium and cerium emissions could be incorporated. Cerium luminescence was detected under 450 nm excitation, while for chromium emission, 270 nm excitation was used. The emission of chromium ions was exceptionally intense, although it was determined that these compounds are doped only by parts per million of Cr^{3+} ions. Typically, the emission maxima of chromium ions are located around 650–750 nm in garnet systems. However, in this case, the emission maximum for chromium is measured to be around 790 nm, caused by re-absorption of Cr^{3+} ions. The main observation of this study is that the switchable emission wavelength in a compound of single phase was obtained, despite the fact that doping with Cr ions was performed in ppm level, causing an intense emission in NIR region.

Keywords: cerium; chromium; garnets; luminescence; NIR; sol-gel synthesis

1. Introduction

In recent years, more and more scientists have focused on the synthesis and development of functional inorganic materials. Two of the main considered groups of such compounds are the inorganic scintillators and phosphors. Inorganic scintillators are widely used in medicine and nuclear physics as the x-ray converter material for CT and others detectors [1–4], while inorganic phosphors are mainly used as material in light-emitting diodes (LEDs) [5–7].

Scintillator and phosphor materials usually consist of a host matrix and activator ions, which, in most cases, are lanthanide or transition metal ions [8]. While the host and activator strategy is the most common, there are other types of materials which possess intrinsic luminescence as well [9,10]. However, lanthanum, promethium, and lutetium are not suitable for such applications. One of the most popular groups of host materials are garnets, which are oxides crystallizing in a cubic structure with space group Ia3d, and which comprise three differently coordinated cation sites, namely dodecahedral, octahedral, and tetrahedral positions [11,12]. The host material used in this study was gadolinium aluminum gallium garnet (GAGG), with the formula $Gd_3Al_2Ga_3O_{12}$. GAGG:Ce garnets exhibited the brightest light yields of 46,000 ph/MeV among other oxide crystalline scintillators [13,14]. Despite the fact that there is a wide variety of dopant ions used in scintillator

production, GAGG is mostly doped just by Ce^{3+} ions [15–18]. There has been a number of studies in recent years about various GAGG:Ce modifications and the resulting changes in luminescence properties. It has been shown that by changing the ratio between Al and Ga in the garnet structure, Ce emission band can be shifted due to the change of crystal field splitting. It is worthy to note that such changes in emission maxima are also present in other garnets doped with cerium ions, such as $Lu_3Al_5O_{12}$:Ce (LuAG:Ce) or $Y_3Al_5O_{12}$:Ce (YAG:Ce) [11,19–21]. Another type of modification for GAGG:Ce is co-doping with various metal cations. Studies have shown that co-doping of GAGG:Ce with mono/di/trivalent cations usually affects the luminescence, i.e., the absorption strength, quantum efficiency, quenching temperature, and decay time [17,22–31]. The change of these parameters depends not only on the dopant type, but also on its concentration. These changes are thought to be caused by the formation of various localized energy levels in the conduction band of the garnet, due to the crystal defects which are formed when Gd^{3+} ions are replaced with cations of different valence c(1+, 2+ or sometimes 4+) [11].

The impact of Cr^{3+} on Ce^{3+} decay time was observed in YAG:Ce,Cr. In this case, the decay time greatly increased with the introduction of high amount (500 ppm) of chromium ions [32]. It was proposed that Cr^{3+} co-doping results in the formation of electron trapping sites with an ideal trap depth for persistent luminescence at room temperature [28,32]. It was also proposed and proven that the same principles can be applied to the GAGG phosphor luminescence where co-doping of Cr^{3+} and Ce^{3+} results in persistent phosphor luminescence of excited Ce^{3+} sites and, in some cases, Cr^{3+} emission in the visible range [28–30,32,33]. In this study, a novel system consisting of GAGG matrix and four additional dopant elements (Ce^{3+}, Cr^{3+}, Mg^{2+} and Ca^{2+}) was investigated. Since the chromium concentration is just in the ppm range, it exhibits not typical chromium emission but the maximum of emission band is shifted towards 790 nm. Such emission in the near infrared region (NIR) range might be useful for plant growth purposes [34]. Additionally, these compounds could be used in the fabrication of high power NIR-LED devices for night vision applications, and biosensors used to measure content of water, fat, sugar, protein of different products [35,36]. Therefore, in turn, these phosphors could be potentially applied in horticulture research experiments. Biological tissue transmits radiation in the range from 650 to 1300 nm (the first biological window), which allows deeper penetration, so such compounds could also be promising candidates in the field of theranostics [37,38].

2. Materials and Methods

Gadolinium aluminum gallium garnet powders were synthesized by the sol-gel method. Cerium concentration was kept at 0.05 mol% for all samples. All synthesized powder compounds are listed in Table 1.

Table 1. Chemical composition of synthesized powders determined by ICP-OES.

Name of Sample	Formula of Sample
Sample 1	GAGG: Ce 0.05%, Ca 100 ppm, Mg 7 ppm, Cr 15 ppm
Sample 2	GAGG: Ce 0.05%, Ca 94 ppm, Mg 8 ppm, Cr 15 ppm
Sample 3	GAGG: Ce 0.05%, Ca 57 ppm, Mg 9 ppm, Cr 15 ppm

For these compounds, Gd_2O_3, Ga_2O_3, $Al(NO_3)_3 \cdot 9H_2O$, $(NH_4)_2Ce(NO_3)_6$, $Ca(NO_3)_2 \cdot 4H_2O$ and $Mg(NO_3)_2 \cdot 6H_2O$ were used as precursors. Firstly, Gd_2O_3 and Ga_2O_3 were dissolved in an excess of concentrated nitric acid at 50 °C. Then, the acid was evaporated and the remaining gel was washed with distilled water 2 or 3 times, followed by further evaporation of added water. An additional 200 mL of water was added after the washing, and $Al(NO_3)_3 \cdot 9H_2O$, $(NH_4)_2Ce(NO_3)_6$, $Ca(NO_3)_2 \cdot 9H_2O$ and $Mg(NO_3)_2 \cdot 9H_2O$ were dissolved. The solution was left under magnetic stirring for 2 h at 50–60 °C. After that, citric acid was added to the solution with a ratio of 3:1 to metal ions, and was left to stir overnight. The solution was evaporated at the same temperature, and the obtained gels were dried at 140 °C for 24 h in the oven. The obtained powders were ground and annealed first at

1000 °C for 2 h in air, with 5 °C/min heating rate. Secondly, the gained powders were heated at 1400 °C for 4 h, with 5 °C/min heating rate.

X-ray diffraction (XRD) measurements of the powders were performed using the Rigaku MiniFlex II X-ray diffractometer (Rigaku Europe SE, Neu-Isenburg, Germany). Powders used for analysis were evenly dispersed on the glass sample holder using ethanol. Then diffraction patterns were recorded in the range of 2θ angles from 15° to 80° for all compounds. Cu K_α radiation (λ = 1.542 Å (Avarage of Cu $K_{\alpha 1}$ = 1.540 and $K_{\alpha 2}$ = 1.544)) was used for the analysis. The measurement parameters were set as follows: current was 15 mA, voltage −30 kV, X-ray detector movement step was 0.010° and dwell time was 5.0 s.

Measurements of emission and excitation: Edinburgh Instruments FLS980 spectrometer (Edinburgh Instruments, Livingston, UK) equipped with double excitation and emission monochromators and 450 W Xe arc lamp (Edinburgh Instruments, Livingston, UK) a cooled (−20 °C) single-photon counting photomultiplier (Hamamatsu R928(Edinburgh Instruments, Livingston, UK)) and mirror optics for powder samples were used for measuring the excitation and emission of the prepared samples. Obtained photoluminescence emission spectra were corrected using a correction file obtained from a tungsten incandescent lamp certified by National Physics Laboratory (NPL), UK. Excitation spectra were corrected by a reference detector [39]. The reflectance spectra from 250 to 600 nm were measured in an integrated (solid) sphere coated with barium sulphate. $BaSO_4$ (99% Sigma-Aldrich, St. Louis, Missouri, United States) was used as a reference material, with excitation and emission gaps of 4 and 0.15 nm, respectively. Each measurement was performed 10 times. Measurements in the range of 120 to 400 nm were performed using a FLS920 fluorescence spectrometer (Edinburgh Instruments, Livingston, UK) with an R-UV excitation monochromator VM504 (Acton Research Corporation, Acton, Massachusetts, USA) and a deuterium lamp (Edinburgh Instruments, Livingston, UK) in a BAM:Eu ($BaMgAl_{10}O_{17}:Eu^{2+}$)-coated integrated sphere. The measuring chamber containing the sample was continuously flushed with dry nitrogen gas to remove water and oxygen, since these molecules show vacuum UV absorption. The photoluminescence decay kinetics were studied for powders and thin films using the FLS980 spectrometer(Edinburgh Instruments, Livingston, UK). 450 nm lasers were used for these measurements [39].

Identification of quantification of Ca, Mg, and Cr in the synthesized species was performed by inductively coupled plasma optical emission spectrometry (ICP-OES) using Perkin-Elmer Optima 7000 DV spectrometer (Perkin-Elmer, Walthman, MA, USA). Sample decomposition procedure was carried out in concentrated nitric acid (HNO_3, Rotipuran® Supra 69% (Roth, Karlsruhe, Germany)) using microwave reaction system Anton Paar Multiwave 3000 (Anton-Paar, Graz, Austria) equipped with XF100 rotor and PTFE liners (Anton-Paar, Graz, Austria). The following program was used for the dissolution of powders: during the first step, the microwave power was linearly increased to 800 W in 15 min and held at this point for the next 20 min. Once the vessels had been fully cooled and depressurized, the obtained clear solutions were quantitatively transferred into volumetric flasks and diluted up to 50 mL with deionized water. Calibration solutions were prepared by an appropriate dilution of the stock standard solutions (single-element ICP standards, 1000 mg/L, (Roth, Karlsruhe, Germany)).

3. Results and Discussion

3.1. X-ray Diffraction Analysis

To determine the phase purity of the powder garnet samples, XRD analysis was performed. Figure 1 shows the diffraction patterns of all synthesized garnets. From the displayed data, it can be concluded that all compounds, regardless of the concentrations of doping elements, possess cubic GAGG structure with Ia-3d space group ((PDF) #00-046-0448). No peaks corresponding to the constituent oxides were observed. However, we can observe a slight shift of the peaks from the PDF card data to smaller 2θ angles. This shift was due to additional doping of the compound with Mg^{2+}, Ca^{2+}, Ce^{3+}, and/or Cr^{3+} ions, due to the ionic radii difference [40]. In summary, from the measured data, it can be concluded that garnets could be doped without the formation of additional impurity phases.

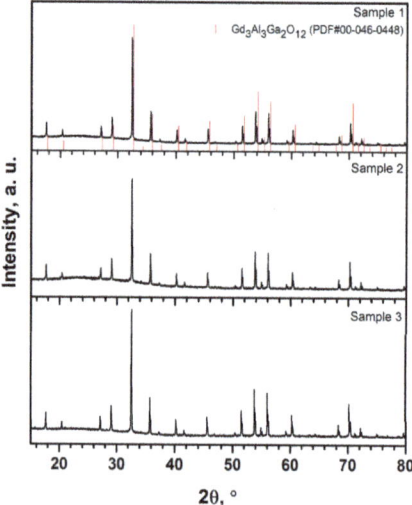

Figure 1. XRD patterns of the GAGG:Ce,Cr,Mg,Ca micropowder samples.

3.2. Luminescence Properties

For the practical application of luminescent materials, they have to emit a rather discreet wavelength light, due to the fact that light interacts differently based on its wavelength. Photons with NIR wavelengths are especially important, and they have deeper tissue penetration and can induce faster plant growth. Additionally, should the possibly arise to switch between NIR and other wavelengths in a single matrix, a new application area could potentially be available. In this case, the luminescence is induced by exciting one of the trivalent ions used as dopants (Ce^{3+}, Cr^{3+}). Upon recording the emission and excitation of the compounds, electronic transitions of both cerium and chromium ions can be observed. The absorption bands that are ascribed to cerium ions are due to inter-configurational $[Xe]4f^1$–$[Xe]5d^1$ transitions, while in the case of Cr^{3+} intraconfigurational $[Ar]3d^3$–$[A]3d^3$ transitions between the crystal-field components, ^4A and ^4T are observed. Due to the allowed nature of these electron transitions, broad emission bands are detected. During these processes, energy is either reemitted or just absorbed, which defines the optical properties of such garnets. Given the possibility to switch between cerium and chromium emission based on the excitation wavelength, as well as the large red shift of chromium emission toward 790 nm, such compounds could potentially be good candidates as multifunctional materials in horticultural and theranostic applications. However, further research for practical applications is still needed to evaluate such possibilities more in-depth.

In order to estimate optical properties of the synthesized materials excitation, emission and reflectance spectra were recorded. The decay times were measured as well. Figures 2 and 3 show the excitation and emission spectra of different GAGG compounds measured at room temperature. In Figure 2, the recorded chromium emission and excitation spectra are displayed. In excitation spectra, the most intense band is attributed to $^8S_{7/2}$–^6I_J transitions of Gd^{3+} ions with the maxima at 270 nm. At 415 and 575 nm, the evidence of excitation process of chromium ions resulting from the transitions between $^4A_1$ to $^4T_1$ and $^4T_2$ orbitals was also observed. Meanwhile, the emission signal detected under 270 nm excitation is caused by the transition from $^4T_2$ to $^4A_1$ orbital. From the spectra given in Figure 2, it can be derived that the compound with the highest content of magnesium

(Sample 3), i.e., 9 ppm, exhibits the most intense emission and excitation bands. The difference in the intensity of excitation and emission could be attributed to the different calcium content in the compounds. Since calcium has a 2+ charge and it is introduced instead of 3+ gadolinium ion, oxygen vacancy defects are most likely created, due to the aliovalent nature of substitution. It is commonly known that such defects reduce the luminescence intensity of most compounds [41,42]. In this case, the compound with the lowest Ca^{2+} amount (Sample 3) has the highest luminescence intensity, whereas Sample 1 and Sample 2 have similar amounts Ca^{2+} ions and show lower emission intensities. It should be noted that chromium is present in ppm fractions, nevertheless, its emission remains very intense. Commonly, chromium emission maximum in garnet matrix is around 650–750 nm, while in this case it was measured to be 790 nm. Such a shift to the red region was previously explained by the re-absorption of Cr^{3+} ions [43]. Studies by other researchers revealed that a high photoluminescence intensity of chromium ions emission was only recorded with 2000 times higher amounts (for example, 3 mol%) of Cr^{3+} in other compounds [44].

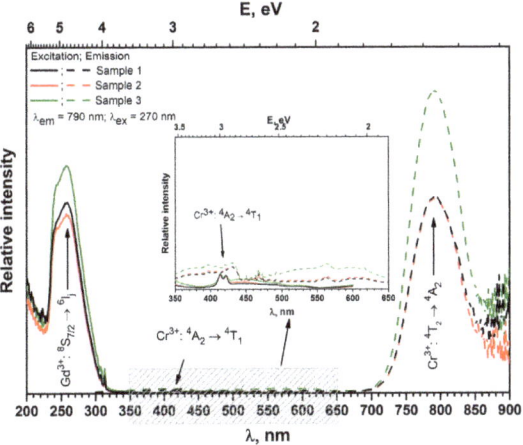

Figure 2. Excitation and emission spectra of GAGG:Ce,Cr,Mg,Ca powder samples (λ_{ex} = 270 nm). Inset represents zoomed in spectral region from 350 to 650 nm.

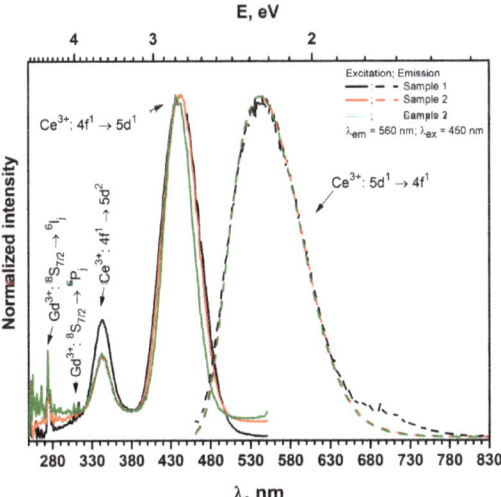

Figure 3. Excitation and emission spectra of GAGG:Ce,Cr,Mg,Ca powder samples (λ_{ex} = 450 nm).

Since these garnets were doped with cerium as well, they also have the characteristic emission attributed to cerium ions. Normalized emission is attributed to cerium emission, and excitation spectra are shown in Figure 3. To measure cerium emission, the compounds were excited under 450 nm wavelength ([Xe]4f^1 to [Xe]5d^1 interconfigurational transitions of Ce^{3+}), which cannot be attributed to chromium. The compounds appear to exhibit cerium-specific fluorescence at a wavelength of 540 nm, which arises from the 5d^1 to 4f^1 electronic transitions. Meanwhile, when investigating the excitation spectra of the compounds in addition to those cerium bands, the peaks contributing to gadolinium ions were observed in the 250–320 nm range [19,44].

Figure 4 shows the reflectance spectra of the compounds. In the reflectance spectra of garnets doped with cerium, two absorption bands can be observed, one at about 350 nm, the another at 400–500 nm. These bands are assigned to the different crystal-field components of the interconfigurational Ce^{3+} [Xe]4f^1 → [Xe]5d^1 transitions. At approximately 275 nm, an absorption peak is observed in all spectra, which is assigned to the Gd^{3+} $^8S_{7/2}$ → ^6I_J interconfigurational transitions. The set of absorption peaks at about 312 nm is assigned to the Gd^{3+} $^8S_{7/2}$ → ^6P_J electron transitions [19,45]. It should be noted that the position of the Ce^{3+} absorption bands does not change upon changing the ions with which the garnet was doped, so it could be stated that doping with magnesium and calcium does not affect the absorption wavelength. However, it is obvious that garnets also have cerium. The absorption band at 360 and 440 nm is classified as a Ce^{3+} electronic transition. In addition, there is no bright absorption band at 600 nm, which is typically observed in Cr^{3+}-containing compounds [44,46–48].

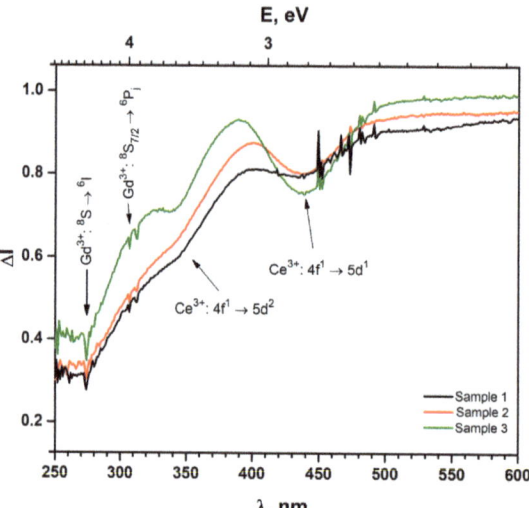

Figure 4. Diffuse reflection spectra of GAGG:Ce,Cr,Mg,Ca powder samples.

To better investigate the absorption of garnets in the UV region, the reflectance spectra of the synthesized gadolinium aluminum gallium garnets powders in the 130–400 nm range were recorded. The reflectance spectra are shown in Figure 5. It can be seen that in the 130–400 nm region, three clear absorption bands are visible: one in the 140–165 nm range, the second in the 190–240 nm range, and the last in 270–280 nm zone. The absorption in the 270–280 nm range can be attributed to Gd^{3+} ^8S → ^6I electron transitions [19]. The observed absorption in the 140–165 nm range can be interpreted as the electron jump into the Ce^{3+} 5d orbital [49,50]. Finally, the region of 200–300 nm can be explained as the band gap absorption of the garnet matrix [50,51].

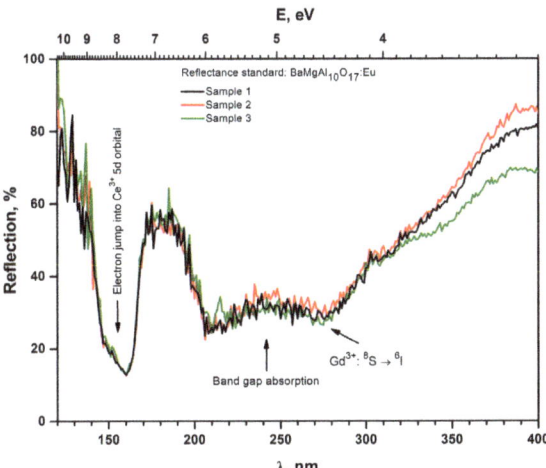

Figure 5. Reflection spectra of GAGG:Ce,Cr,Mg,Ca powder samples in the UV region.

The photoluminescence decay curves of the chromium emission are shown in Figure 6. All calculated values of the decay times are listed in Table 2. As can be seen from Table 2, all compounds have almost the same decay times, which are between 6.0 and 5.9 ms. It can be seen that the samples have emission decay times showing the characteristics of Cr^{3+} ions, because characteristic decay times for Ce^{3+} ions are much shorter and reach about 50–60 ns [5,52]. Cr^{3+} ions electron transition are considered forbidden, which usually results in longer decay times in a rage of milliseconds [42,53]. However, for the Ce^{3+} ions, electron transitions are allowed, and the decay time is in the range of nanoseconds [5]. The obtained values are of chromium, since the 270 nm excitation wavelength was used. While there have been previous reports on the effect of calcium and magnesium doping in garnet matrix on decay times [17,54–56], in this case, no effect was observed, potentially due to the drastically smaller amounts to aforementioned ions.

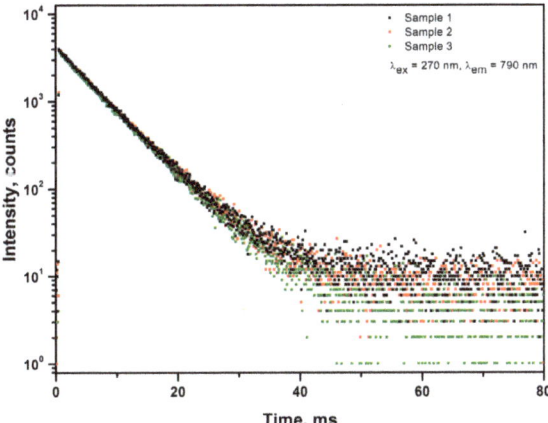

Figure 6. Decay curves of the Cr^{3+} emission of the GAGG:Ce,Cr,Mg,Ca powder samples upon 270 nm excitation.

Table 2. Synthesized powder samples.

Sample	Decay Time (ms)
GAGG: Ce 0.05%, Ca 100 ppm, Mg 7 ppm, Cr 15 ppm (Sample 1)	5.9
GAGG: Ce 0.05%, Ca 94 ppm, Mg 8 ppm, Cr 15 ppm (Sample 2)	6
GAGG: Ce 0.05%, Ca 57 ppm, Mg 9 ppm, Cr 15 ppm (Sample 3)	6

4. Conclusions

Gadolinium aluminum gallium garnets doped with 0.05% cerium, chromium, magnesium, and calcium were synthesized by the sol-gel method whereby obtained garnet type samples are of single phase. Measurement of the excitation and emission spectra of these compounds revealed that extremely low levels of Cr^{3+} ions, i.e., only at a level of 15 ppm, caused intense emission in the NIR region. At the same time, but under different excitation wavelengths, a typical emission spectrum of Ce^{3+} was also observed, which proved that the single compound might emit in very broad range (from 470 to 850 nm). The most intense emission was located at a wavelength of 790 nm. The obtained compounds exhibit promising luminescence properties to illuminate plants and promote their growth. Since luminescence covered the first biological window and chromium ions exhibited a long specific decay time in the range of 6 ms, the synthesized compounds also demonstrate characteristics required for bio-imaging purposes.

Author Contributions: R.S. and G.I. conceived and planned the experiments and wrote the manuscript with support from other co-authors, G.L. and G.I. synthesized powders, R.S. supervised the project, A.Z. performed ICP measurements of powders, G.L. performed XRD measurements, D.E., G.L. and T.J. performed luminescence measurements, and T.J. analyzed luminescence properties. All authors provided critical feedback and helped shape the research, analysis and manuscript. All authors have read and agreed to the published version of the manuscript.

Funding: This research received no external funding.

Institutional Review Board Statement: Not applicable.

Informed Consent Statement: Not applicable.

Conflicts of Interest: The authors declare no conflict of interest.

References

1. Paper, C.; Cherepy, N.; Livermore, L.; Seeley, Z.M.; Livermore, L.; Beck, P.; Livermore, L.; Hunter, S.L.; Livermore, L. High Energy Resolution Transparent Ceramic Garnet Scintillators. *Int. Soc. Opt. Photonics* **2014**, *2*. [CrossRef]
2. Sakthong, O.; Chewpraditkul, W.; Wanarak, C.; Kamada, K. Nuclear Instruments and Methods in Physics Research A Scintillation Properties of $Gd_3Al_2Ga_3O_{12}$:Ce_3þ Single Crystal Scintillators. *Nucl. Inst. Methods Phys. Res. A* **2014**, *751*, 1–5. [CrossRef]
3. Belli, P.; Bernabei, R.; Cerulli, R.; Dai, C.J.; Danevich, F.A.; Incicchitti, A.; Kobychev, V.V.; Ponkratenko, O.A.; Prosperi, D.; Tretyak, V.I.; et al. Performances of a CeF_3 Crystal Scintillator and Its Application to the Search for Rare Processes. *Nucl. Instrum. Methods Phys. Res. Sect. A Accel. Spectrometers Detect. Assoc. Equip.* **2003**, *498*, 352–361. [CrossRef]
4. Van Eijk, C.W.E. *Scintillator-Based Detectors*; Elsevier B.V.: Amsterdam, The Netherlands, 2014; Volume 8. [CrossRef]
5. Bachmann, V.; Ronda, C.; Meijerink, A. Temperature Quenching of Yellow Ce^{3+} Luminescence in YAG:Ce. *Chem. Mater.* **2009**, *21*, 2077–2084. [CrossRef]
6. Khaidukov, N.; Zorenko, T.; Iskaliyeva, A.; Paprocki, K.; Batentschuk, M.; Osvet, A.; Van Deun, R.; Zhydaczevskii, Y.; Suchocki, A.; Zorenko, Y. Synthesis and Luminescent Properties of Prospective Ce^{3+} Doped Silicate Garnet Phosphors for White LED Converters. *J. Lumin.* **2017**, *192*, 328–336. [CrossRef]
7. Ueda, J.; Dorenbos, P.; Bos, A.J.J.; Meijerink, A.; Tanabe, S. Insight into the Thermal Quenching Mechanism for $Y_3Al_5O_{12}$:Ce^{3+} through Thermoluminescence Excitation Spectroscopy. *J. Phys. Chem. C* **2015**, *119*, 25003–25008. [CrossRef]
8. Auffray, E.; Korjik, M.; Lucchini, M.T.; Nargelas, S.; Sidletskiy, O.; Tamulaitis, G.; Tratsiak, Y.; Vaitkevičius, A. Free Carrier Absorption in Self-Activated $PbWO_4$ and Ce-Doped $Y_3(Al_{0.25}Ga_{0.75})_3O_{12}$ and $Gd_3Al_2Ga_3O_{12}$ Garnet Scintillators. *Opt. Mater.* **2016**, *58*, 461–465. [CrossRef]
9. Seminko, V.; Maksimchuk, P.; Bespalova, I.; Masalov, A.; Viagin, O.; Okrushko, E.; Kononets, N.; Malyukin, Y. Defect and Intrinsic Luminescence of CeO_2 Nanocrystals. *Phys. Status Solidi* **2017**, *254*, 1600488. [CrossRef]

10. Zatsepin, D.A.; Boukhvalov, D.W.; Zatsepin, A.F.; Kuznetsova, Y.A.; Mashkovtsev, M.A.; Rychkov, V.N.; Shur, V.Y.; Esin, A.A.; Kurmaev, E.Z. Electronic Structure, Charge Transfer, and Intrinsic Luminescence of Gadolinium Oxide Nanoparticles: Experiment and Theory. *Appl. Surf. Sci.* **2018**, *436*, 697–707. [CrossRef]
11. Kanai, T.; Satoh, M.; Miura, I. Characteristics of a Nonstoichiometric $Gd^{3+\delta}(Al,Ga)_{5-\Delta}O_{12}$:Ce Garnet Scintillator. *J. Am. Ceram. Soc.* **2008**, *91*, 456–462. [CrossRef]
12. Geller, S. Magnetic Interactions and Distribution of Ions in the Garnets. *J. Appl. Phys.* **1960**, *30*. [CrossRef]
13. Yanagida, T.; Kamada, K.; Fujimoto, Y.; Yagi, H.; Yanagitani, T. Comparative Study of Ceramic and Single Crystal Ce:GAGG Scintillator. *Opt. Mater.* **2013**, *35*, 2480–2485. [CrossRef]
14. Fedorov, A.; Gurinovich, V.; Guzov, V.; Dosovitskiy, G.; Korzhik, M.; Kozhemyakin, V.; Lopatik, A.; Kozlov, D.; Mechinsky, V.; Retivov, V. Sensitivity of GAGG Based Scintillation Neutron Detector with SiPM Readout. *Nucl. Eng. Technol.* **2020**, *52*, 2306–2312. [CrossRef]
15. Kitaura, M.; Sato, A.; Kamada, K.; Kurosawa, S.; Ohnishi, A.; Sasaki, M.; Hara, K. Photoluminescence Studies on Energy Transfer Processes In. *Opt. Mater.* **2015**, *41*, 45–48. [CrossRef]
16. Kitaura, M.; Sato, A.; Kamada, K.; Ohnishi, A.; Sasaki, M. Phosphorescence of Ce-Doped $Gd_3Al_2Ga_3O_{12}$ Crystals Studied Using Luminescence Spectroscopy. *J. Appl. Phys.* **2014**, *115*, 10–18. [CrossRef]
17. Lucchini, M.T.; Gundacker, S.; Lecoq, P.; Benaglia, A.; Nikl, M.; Kamada, K.; Yoshikawa, A.; Auffray, E. Timing Capabilities of Garnet Crystals for Detection of High Energy Charged Particles. *Nucl. Instruments Methods Phys. Res. Sect. A Accel. Spectrometers Detect. Assoc. Equip.* **2017**, *852*, 1–9. [CrossRef]
18. Gundacker, S.; Turtos, R.M.; Auffray, E.; Lecoq, P. Precise Rise and Decay Time Measurements of Inorganic Scintillators by Means of X-Ray and 511 KeV Excitation. *Nucl. Instrum. Methods Phys. Res. Sect. A Accel. Spectrometers Detect. Assoc. Equip.* **2018**, *891*, 42–52. [CrossRef]
19. Ogieg, J.M.; Katelnikovas, A.; Zych, A.; Ju, T.; Meijerink, A.; Ronda, C.R. Luminescence and Luminescence Quenching in $Gd_3(Ga,Al)_5O_{12}$ Scintillators Doped with Ce^{3+}. *J. Phys. Chem. A* **2013**, *117*, 2479–2484. [CrossRef] [PubMed]
20. Ueda, J.; Tanabe, S.; Nakanishi, T. Analysis of Ce^{3+} Luminescence Quenching in Solid Solutions between $Y_3Al_5O_{12}$ and $Y_3Ga_5O_{12}$ by Temperature Dependence of Photoconductivity Measurement. *J. Appl. Phys.* **2011**, *110*, 053102. [CrossRef]
21. Pan, Y.; Wu, M.; Su, Q. Tailored Photoluminescence of YAG:Ce Phosphor through Various Methods. *J. Phys. Chem. Solids* **2004**, *65*, 845–850. [CrossRef]
22. Yoshino, M.; Kamada, K.; Kochurikhin, V.V.; Ivanov, M.; Nikl, M.; Okumura, S.; Yamamoto, S.; Yeom, J.Y.; Shoji, Y.; Kurosawa, S.; et al. Li^+, Na^+ and K^+ Co-Doping Effects on Scintillation Properties of $Ce:Gd_3Ga_3Al_2O_{12}$ Single Crystals. *J. Cryst. Growth* **2018**, *491*, 1–5. [CrossRef]
23. Kobayashi, T.; Yamamoto, S.; Yeom, J.; Kamada, K.; Yoshikawa, A. Development of High Resolution Phoswich Depth-of-Interaction Block Detectors Utilizing Mg Co-Doped New Scintillators. *Nucl. Inst. Methods Phys. Res. A* **2017**. [CrossRef]
24. Meng, F.; Koschan, M.; Wu, Y.; Melcher, C.L. Relationship between Ca^{2+} Concentration and the Properties of Codoped $Gd_3Ga_3Al_2O_{12}$:Ce Scintillators. *Nucl. Instrum. Methods Phys. Res. Sect. A Accel. Spectrometers Detect. Assoc. Equip.* **2015**, *797*, 138–143. [CrossRef]
25. Kamada, K.; Shoji, Y.; Kochurikhin, V.V.; Nagura, A.; Okumura, S.; Yamamoto, S.; Yeom, J.Y.; Kurosawa, S.; Pejchal, J.; Yokota, Y.; et al. Single Crystal Growth of $Ce:Gd_3(Ga,Al)_5O_{12}$ with Various Mg Concentration and Their Scintillation Properties. *J. Cryst. Growth* **2017**, *468*, 407–410. [CrossRef]
26. Lucchini, M.T.; Babin, V.; Bohacek, P.; Gundacker, S.; Kamada, K.; Nikl, M.; Petrosyan, A.; Yoshikawa, A.; Auffray, E. Effect of Mg^{2+} Ions Co-Doping on Timing Performance and Radiation Tolerance of Cerium Doped $Gd_3Al_2Ga_3O_{12}$ Crystals. *Nucl. Instrum. Methods Phys. Res. Sect. A Accel. Spectrometers Detect. Assoc. Equip.* **2016**, *816*, 176–183. [CrossRef]
27. Auffray, E.; Augulis, R.; Fedorov, A.; Dosovitskiy, G.; Grigorjeva, L.; Gulbinas, V.; Koschan, M.; Lucchini, M.; Melcher, C.; Nargelas, S.; et al. Excitation Transfer Engineering in Ce-Doped Oxide Crystalline Scintillators by Codoping with Alkali-Earth Ions. *Phys. Status Solidi Appl. Mater. Sci.* **2018**, *215*, 1–10. [CrossRef]
28. Ueda, J.; Kuroishi, K.; Tanabe, S. Yellow Persistent Luminescence in Ce^{3+}-Cr^{3+}-Codoped Gadolinium Aluminum Gallium Garnet Transparent Ceramics after Blue-Light Excitation. *Appl. Phys. Express* **2014**, *7*. [CrossRef]
29. Asami, K.; Ueda, J.; Tanabe, S. Trap Depth and Color Variation of Ce^{3+}-Cr^{3+} Co-Doped $Gd_3(Al,Ga)_5O_{12}$ Garnet Persistent Phosphors. *Opt. Mater.* **2016**, *62*, 171–175. [CrossRef]
30. Gotoh, T.; Jeem, M.; Zhang, L.; Okinaka, N.; Watanabe, S. Synthesis of Yellow Persistent Phosphor Garnet by Mixed Fuel Solution Combustion Synthesis and Its Characteristic. *J. Phys. Chem. Solids* **2020**, *142*, 109436. [CrossRef]
31. Tamulaitis, G.; Vaitkevičius, A.; Nargelas, S.; Augulis, R.; Gulbinas, V.; Bohacek, P.; Nikl, M.; Borisevich, A.; Fedorov, A.; Korjik, M.; et al. Subpicosecond Luminescence Rise Time in Magnesium Codoped GAGG:Ce Scintillator. *Nucl. Instruments Methods Phys. Res. Sect. A Accel. Spectrometers Detect. Assoc. Equip.* **2017**, *870*, 25–29. [CrossRef]
32. Ueda, J.; Kuroishi, K.; Tanabe, S. Bright Persistent Ceramic Phosphors of Ce^{3+}-Cr^{3+}-Codoped Garnet Able to Store by Blue Light. *Appl. Phys. Lett.* **2014**, *104*, 5–9. [CrossRef]
33. Blasse, G.; Grabmaier, B.C.; Ostertag, M. The Afterglow Mechanism of Chromium-Doped Gadolinium Gallium Garnet. *J. Alloys Compd.* **1993**, *200*, 17–18. [CrossRef]
34. Yamada, A.; Tanigawa, T.; Suyama, T.; Matsuno, T.; Kunitake, T. Red:Far-Red Light Ratio and Far-Red Light Integral Promote or Retard Growth and Flowering in Eustoma Grandiflorum (Raf.) Shinn. *Sci. Hortic.* **2009**, *120*, 101–106. [CrossRef]

35. Jia, Z.; Yuan, C.; Liu, Y.; Wang, X.J.; Sun, P.; Wang, L.; Jiang, H.; Jiang, J. Strategies to Approach High Performance in Cr^{3+}-Doped Phosphors for High-Power NIR-LED Light Sources. *Light Sci. Appl.* **2020**, *9*, 2047–7538. [CrossRef]
36. He, S.; Zhang, L.; Wu, H.; Wu, H.; Pan, G.; Hao, Z.; Zhang, X.; Zhang, L.; Zhang, H.; Zhang, J. Efficient Super Broadband NIR $Ca_2LuZr_2Al_3O_{12}:Cr^{3+},Yb^{3+}$ Garnet Phosphor for Pc-LED Light Source toward NIR Spectroscopy Applications. *Adv. Opt. Mater.* **2020**, *8*, 1901684. [CrossRef]
37. Ferrauto, G.; Carniato, F.; Di Gregorio, E.; Botta, M.; Tei, L. Photoacoustic Ratiometric Assessment of Mitoxantrone Release from Theranostic ICG-Conjugated Mesoporous Silica Nanoparticles. *Nanoscale* **2019**, *11*, 18031–18036. [CrossRef] [PubMed]
38. Cao, R.; Shi, Z.; Quan, G.; Chen, T.; Guo, S.; Hu, Z.; Liu, P. Preparation and Luminescence Properties of $Li_2MgZrO_4:Mn^{4+}$ Red Phosphor for Plant Growth. *J. Lumin.* **2017**, *188*, 577–581. [CrossRef]
39. Inkrataite, G.; Zabiliute-Karaliune, A.; Aglinskaite, J.; Vitta, P.; Kristinaityte, K.; Marsalka, A.; Skaudzius, R. Study of YAG:Ce and Polymer Composite Properties for Application in LED Devices. *Chempluschem* **2020**, *85*, 1504–1510. [CrossRef]
40. Chen, X.; Qin, H.; Zhang, Y.; Jiang, J.; Wu, Y.; Jiang, H. Effects of Ga Substitution for Al on the Fabrication and Optical Properties of Transparent Ce:GAGG-Based Ceramics. *J. Eur. Ceram. Soc.* **2017**, *37*, 4109–4114. [CrossRef]
41. Panatarani, C.; Faizal, F.; Florena, F.F.; Jumhur, D.; Made Joni, I. The Effects of Divalent and Trivalent Dopants on the Luminescence Properties of ZnO Fine Particle with Oxygen Vacancies. *Adv. Powder Technol.* **2020**, *31*, 2605–2612. [CrossRef]
42. Dickens, P.T.; Haven, D.T.; Friedrich, S.; Lynn, K.G. Scintillation Properties and Increased Vacancy Formation in Cerium and Calcium Co-Doped Yttrium Aluminum Garnet. *J. Cryst. Growth* **2019**, *507*, 16–22. [CrossRef]
43. Mao, M.; Zhou, T.; Zeng, H.; Wang, L.; Huang, F.; Tang, X.; Xie, R.J. Broadband Near-Infrared (NIR) Emission Realized by the Crystal-Field Engineering of $Y_{3-x}Ca_xAl_{5-x}Si_xO_{12}:Cr^{3+}$ (x = 0-2.0) Garnet Phosphors. *J. Mater. Chem. C* **2020**, *8*, 1981–1988. [CrossRef]
44. Butkute, S.; Gaigalas, E.; Beganskiene, A.; Ivanauskas, F.; Ramanauskas, R.; Kareiva, A. Sol-Gel Combustion Synthesis of High-Quality Chromium-Doped Mixed-Metal Garnets $Y_3Ga_5O_{12}$ and $Gd_3Sc_2Ga_3O_{12}$. *J. Alloys Compd.* **2018**, *739*, 504–509. [CrossRef]
45. Chen, L.; Chen, X.; Liu, F.; Chen, H.; Wang, H.; Zhao, E.; Jiang, Y.; Chan, T.S.; Wang, C.H.; Zhang, W.; et al. Charge Deformation and Orbital Hybridization: Intrinsic Mechanisms on Tunable Chromaticity of $Y_3Al_5O_{12}:Ce^{3+}$ Luminescence by Doping Gd^{3+} for Warm White LEDs. *Sci. Rep.* **2015**, *5*, 1–17. [CrossRef]
46. Yanagida, T.; Fujimoto, Y.; Yamaji, A.; Kawaguchi, N.; Kamada, K.; Totsuka, D.; Fukuda, K.; Yamanoi, K.; Nishi, R.; Kurosawa, S.; et al. Study of the Correlation of Scintillation Decay and Emission Wavelength. *Radiat. Meas.* **2013**, *55*, 99–102. [CrossRef]
47. Malysa, B.; Meijerink, A.; Jüstel, T. Temperature Dependent Luminescence Cr^{3+}-Doped $GdAl_3(BO_3)_4$ and $YAl_3(BO_3)_4$. *J. Lumin.* **2016**, *171*, 246–253. [CrossRef]
48. Zhou, X.; Luo, X.; Wu, B.; Jiang, S.; Li, L.; Luo, X.; Pang, Y. The Broad Emission at 785 Nm in $YAG:Ce^{3+}$, Cr^{3+} Phosphor. *Spectrochim. Acta Part A Mol. Biomol. Spectrosc.* **2018**, *190*, 76–80. [CrossRef]
49. Yanagida, T.; Fujimoto, Y.; Koshimizu, M.; Watanabe, K.; Sato, H.; Yagi, H.; Yanagitani, T. Positive Hysteresis of Ce-Doped GAGG Scintillator. In *Optical Materials*; Elsevier B.V.: Amsterdam, The Netherlands, 2014; Volume 36, pp. 2016–2019. [CrossRef]
50. Liu, W.-R.; Lin, C.C.; Chiu, Y.-C.; Yeh, Y.-T.; Jang, S.-M.; Liu, R.-S.; Cheng, B.-M. Versatile Phosphors $BaY_2Si_3O_{10}$:RE (RE = Ce^{3+}, Tb^{3+}, Eu^{3+}) for Light-Emitting Diodes. *Opt. Express* **2009**, *17*, 18103. [CrossRef]
51. Spassky, D.; Kozlova, N.; Zabelina, E.; Kasimova, V.; Krutyak, N.; Ukhanova, A.; Morozov, V.A.; Morozov, A.V.; Buzanov, O.; Chernenko, K.; et al. Influence of the Sc Cation Substituent on the Structural Properties and Energy Transfer Processes in GAGG:Ce Crystals. *CrystEngComm* **2020**, *22*, 2621–2631. [CrossRef]
52. Katelnikovas, A.; Bettentrup, H.; Dutczak, D.; Kareiva, A.; Jüstel, T. On the Correlation between the Composition of Pr3 Doped Garnet Type Materials and Their Photoluminescence Properties. *J. Lumin.* **2011**, *131*, 2754–2761. [CrossRef]
53. Zhou, D.; Tao, L.; Yu, Z.; Jiao, J.; Xu, W. Efficient Chromium Ion Passivated $CsPbCl_3$:Mn Perovskite Quantum Dots for Photon Energy Conversion in Perovskite Solar Cells. *J. Mater. Chem. C* **2020**, *8*, 12323–12329. [CrossRef]
54. Malysa, B.; Meijerink, A.; Jüstel, T. Temperature Dependent Cr^{3+} Photoluminescence in Garnets of the Type $X_3Sc_2Ga_3O_{12}$ (X = Lu, Y, Gd, La). *J. Lumin.* **2018**, *202*, 523–531. [CrossRef]
55. Dantelle, G.; Boulon, G.; Guyot, Y.; Testemale, D.; Guzik, M.; Kurosawa, S.; Kamada, K.; Yoshikawa, A. Research on Efficient Fast Scintillators: Evidence and X-Ray Absorption Near Edge Spectroscopy Characterization of Ce^{4+} in $Ce^{3+,}$ Mg^{2+}-Co-Doped $Gd_3Al_2Ga_3O_{12}$ Garnet Crystal. *Phys. Status Solidi Basic Res.* **2020**, *257*, 1900510. [CrossRef]
56. Lucchini, M.T.; Buganov, O.; Auffray, E.; Bohacek, P.; Korjik, M.; Kozlov, D.; Nargelas, S.; Nikl, M.; Tikhomirov, S.; Tamulaitis, G.; et al. Measurement of Non-Equilibrium Carriers Dynamics in Ce-Doped YAG, LuAG and GAGG Crystals with and without Mg-Codoping. *J. Lumin.* **2018**, *194*, 1–7. [CrossRef]

Article

Synthesis of Carbon-Supported MnO₂ Nanocomposites for Supercapacitors Application

Jolita Jablonskiene *, Dijana Simkunaite, Jurate Vaiciuniene, Giedrius Stalnionis, Audrius Drabavicius, Vitalija Jasulaitiene, Vidas Pakstas, Loreta Tamasauskaite-Tamasiunaite * and Eugenijus Norkus

Center for Physical Sciences and Technology, Sauletekio Ave. 3, LT-10257 Vilnius, Lithuania; dijana.simkunaite@ftmc.lt (D.S.); jurate.vaiciuniene@ftmc.lt (J.V.); giedrius.stalnionis@ftmc.lt (G.S.); audrius.drabavicius@ftmc.lt (A.D.); vitalija.jasulaitiene@ftmc.lt (V.J.); vidas.pakstas@ftmc.lt (V.P.); eugenijus.norkus@ftmc.lt (E.N.)
* Correspondence: jolita.jablonskiene@ftmc.lt (J.J.); loreta.tamasauskaite@ftmc.lt (L.T.-T.)

Abstract: In this study, carbon-supported MnO₂ nanocomposites have been prepared using the microwave-assisted heating method followed by two different approaches. The MnO₂/C nanocomposite, labeled as sample S1, was prepared directly by the microwave-assisted synthesis of mixed KMnO₄ and carbon powder components. Meanwhile, the other MnO₂/C nanocomposite sample labeled as S2 was prepared indirectly via a two-step procedure that involves the microwave-assisted synthesis of mixed KMnO₄ and MnSO₄ components to generate MnO₂ and subsequent secondary microwave heating of synthesized MnO₂ species coupled with graphite powder. Field emission scanning electron microscopy (FE-SEM), transmission electron microscopy (TEM), X-ray photoelectron spectroscopy (XPS), and inductively coupled plasma optical emission spectroscopy have been used for characterization of MnO₂/C nanocomposites morphology, structure, and composition. The electrochemical performance of nanocomposites has been investigated using cyclic voltammetry and galvanostatic charge/discharge measurements in a 1 M Na₂SO₄ solution. The MnO₂/C nanocomposite, prepared indirectly via a two-step procedure, displays substantially enhanced electrochemical characteristics. The high specific capacitance of 980.7 F g^{-1} has been achieved from cyclic voltammetry measurements, whereas specific capacitance of 949.3 F g^{-1} at 1 A g^{-1} has been obtained from galvanostatic charge/discharge test for sample S2. In addition, the specific capacitance retention was 93% after 100 cycles at 20 A g^{-1}, indicating good electrochemical stability.

Keywords: supercapacitors; microwave synthesis; nanocomposites; MnO₂

Citation: Jablonskiene, J.; Simkunaite, D.; Vaiciuniene, J.; Stalnionis, G.; Drabavicius, A.; Jasulaitiene, V.; Pakstas, V.; Tamasauskaite-Tamasiunaite, L.; Norkus, E. Synthesis of Carbon-Supported MnO₂ Nanocomposites for Supercapacitors Application. *Crystals* **2021**, *11*, 784. https://doi.org/10.3390/cryst11070784

Academic Editor: Fabrizio Pirri

Received: 11 May 2021
Accepted: 1 July 2021
Published: 5 July 2021

Publisher's Note: MDPI stays neutral with regard to jurisdictional claims in published maps and institutional affiliations.

Copyright: © 2021 by the authors. Licensee MDPI, Basel, Switzerland. This article is an open access article distributed under the terms and conditions of the Creative Commons Attribution (CC BY) license (https://creativecommons.org/licenses/by/4.0/).

1. Introduction

The development of high-performance, environmentally friendly, flexible, light and inexpensive energy storage devices has become one of the most significant worldwide concerns over the past few decades [1–4]. In this regard, supercapacitors (SCs) are widely viewed as potential candidates for next-generation energy storage devices [5,6]. They are of particular interest for high power capability, cyclic stability, safe and simple operation principle, and speedy charge dynamics compared to the other storage devices [7,8].

The decisive role in the fabrication of efficient SCs is directly related to the intrinsic properties of the electrode material used [2,9]. Various new advanced nanostructured materials or their hybrid combinations of two or more components are being looked at. Currently, transition metal oxides (TMOs) coupled mainly with high-surface-area carbon-based materials have been allowed to achieve various hybrid SCs systems, which have superior characteristics of power and energy densities compared to those values obtained at each system separately [10,11]. Among the preferential TMO materials for SCs, a series of manganese oxide-based hybrids were developed recently [12–16]. Exceptional attention has been focused on MnO₂ for its relatively low cost, environmental friendliness, natural abundance, multiple oxidation states of Mn, wide operating voltage range (0–1.00 V vs.

NHE in the neutral electrolyte), and high theoretical specific capacitance (C_s) close to 1370 F g^{-1} [17,18].

However, regardless of the above-mentioned superb characteristics, the main drawback for MnO$_2$ widespread application is relatively poor electronic (10^{-5}–10^{-6} S cm^{-1}) and ionic (10^{-13} S cm^{-1}) conductivity [8,15]. The experimentally achieved actual capacitance value is far below the theoretically predicted value and depends strongly on the mass loading of MnO$_2$. Typically it decreases rapidly with the increase in the MnO$_2$ mass [19]. To improve the capacitive performance of MnO$_2$, several strategies have been proposed. Nanostructured MnO$_2$-based electrodes with various morphologies have a high specific surface area and a large surface-to-volume ratio for more effective contact with electrolyte ions, such as mesoporous MnO$_2$ nanotubes/nanosheets [20], nanowires [21], or flower-like, urchinlike, and nano rodlike structures that have been developed. Multiple-phase heterostructures for high-capacitance electrodes have been created as well [22]. Two high-capacitance crystal phases of MnO$_2$, namely α-MnO$_2$ nanowires and δ-MnO$_2$ ultrathin nanoflakes, have been combined and generated a self-branch heterostructure with a high C_s value of 178 F g^{-1} at 5 mV s^{-1} [22].

To enhance the electric conductivity of MnO$_2$, the incorporation of conductive metals including Au, Al, Cu, Fe, Mg, Co [23–31] able to act as electron donors have been applied. Changes in electron structure by foreign heteroatoms resulted in the improved capacitive performance of MnO$_2$ and revealed that, for example, Cu-doped δ-MnO$_2$ film delivered the maximum C_s value as high as 296 F g^{-1} at 1 A g^{-1} [26]; for Fe-doped MnO$_2$ nanostructures, this value was of 267.0 F g^{-1} even under a high mass loading of 5 mg cm^{-2} [28]. In the presence of Co, the achieved C_s value was of 350 F g^{-1} at a current density of 0.1 A g^{-1} [29]. The Al-doped MnO$_2$ demonstrated a high mass and areal specific capacitance of 213 F g^{-1} and 146 F cm^{-2}, respectively, at 0.1 A g^{-1} [25]. Meanwhile, Au-doped MnO$_2$ showed a high C_s value of 626 F g^{-1} at 5 mV s^{-1} [24].

However, the most effective and currently most widely used way to improve C_s of MnO$_2$-based electrodes is the deposition of thin films of latter materials on highly conductive and large surface areas containing materials, such as carbon-based substrates, including activated carbon, carbon nanotubes (CNTs), graphene, carbon fiber, or graphitic carbon. Carbon-based materials are the most widely used because of their physical and chemical properties, including low cost, variety of forms, low effort of processing, relatively inert electrochemistry, controllable porosity, and numerous electrocatalytic active sites for a variety of redox reactions [32–34] However, the performance of carbon-based substrates has some limitations related to the insufficient penetration of ions on the inert surface. Therefore, nanohybrids from two or more materials have been developed to overcome such limitations and gained special attention due to synergetic effects in enhancing the surface and electron donor properties. S.V. Prabhakar Vattikuti et al. reported 1D/2Dcarbon-CuO-graphitic carbon nitride (C/CuO@g-C$_3$N$_4$) ternary heterostructure that showed a better specific capacitance of 247.2 F g^{-1} compared with the pristine g-C$_3$N$_4$ of 83.7 F g^{-1}, at the same time possessing good stability, with 92.1% of the initial capacitance remaining even after 6000 cycles [35]. Newly designed nanohybrids with Bi$_2$S$_3$ nanorod core@ amorphous carbon shell heterostructure C@Bi$_2$S$_3$ displayed a high specific capacity of 333.43 F g^{-1} at a current density of 1 A g^{-1} and outperformed that of pristine Bi$_2$S$_3$ of 124.24 F g^{-1}, due to well-defined cross linkages between the Bi$_2$S$_3$ core and carbon shell [36]. The carbon layer was supposed to bind efficiently with Bi$_2$S$_3$ nanorods, and thus improve electrical contact with the current collector, confirming the more active carbon participation in the charge/discharge reaction process. These highly porous structures allow the free permeation of electrolyte ensuring rapid movement of ions. Further on, a novel Na$_2$Ti$_3$O$_7$/single-walled carbon nanotubes SWCNTs nanostructure electrode material due to high surface area, enriched interfacial conductivity, abundant active edge sites, and mesoporous nature demonstrated a capacity of 576.01 F g^{-1}, at 0.8 A g^{-1}, with cycling stability featuring 91.43% retaining of capacitance after 5000 cycles [37].

Recently, a synergy of such carbon-based materials, featuring excellent conductivity and ultrastability, with MnO$_2$ substances having less lower conductivity but larger electrochemical capacitance, has allowed overcoming limitations of each material separately by making full use of their advantages due to the synergistic effects between those two types of SCs materials [38–44]. Y. Ping et al. produced the hierarchically porous CJE/MnO$_2$ composite with a large specific capacitance of 283 F g^{-1} at 1 A g^{-1}, which was on account of high specific surface area (1283 m^2 g^{-1}) and abundant active sites for pseudocapacitance, that particularly resulted from the introduction of MnO$_2$ [38]. Meanwhile, highly loaded MnOx of 7.02 mg cm^{-2}, electrodeposited on conductive carbon cloth allowed achieving excellent rate capability due to the dual-tuning effect and showed specific capacitance of 161.2 F g^{-1} (1.13 F cm^{-2}) at a high current density of 20 mA cm^{-2} [39]. Recently, an MnO$_2$ nanowires/graphenated CNTs composite was grown in situ on 316 L stainless steel and exhibited a high capacitance of 495.2 mF cm^{-2} (615.6 F g^{-1}) at a current density of 0.5 mA cm^{-2} and 95% capacity retention after 5000 cycles due to the synergistic effects of the high conductivity of graphenated CNTs and high pseudocapacitance of MnO$_2$ nanowires [42].

Bearing in mind that the structure of the electrode material directly affects the electrochemical properties of the electrode and simultaneously determines the performance of SCs, different methods have been tested to develop high capacitive MnO$_2$-based electrode materials. Among them are such methods as hydrothermal synthesis [20,45,46], electrochemical deposition [42,47,48], electrochemical exfoliation [48], electrospinning [49], chemical coprecipitation [50], or even those, using templates [40]. Particular attention has been focused on the simple, fast, cost-effective, and reliable microwave-assisted approach. This method has several advantages that count the possibility to get great gain in energy savings and enhanced fabrication of homogeneous materials since they do not need expensive equipment or complicated procedures; microwave reactions take less time compared to conventional methods and overtake all the substance uniformly, providing uniform particle-size distribution in the sample. Recently, this approach was successfully introduced to synthesize MnO$_2$ materials that proved themselves for possible use in high-performance supercapacitor applications [51–57].

In this study, the carbon-supported MnO$_2$ nanocomposites (MnO$_2$/C) were prepared by the rapid and simple microwave-assisted heating method by employing manganese(II) sulfate (MnSO$_4$) or potassium permanganate (KMnO$_4$) and carbon powder as the microwave absorbing material. The electrochemical properties of the prepared MnO$_2$/C nanocomposites have been studied to evaluate the possibility of using these nanocomposites as potential supercapacitor electrode materials.

2. Materials and Methods

Graphite powder, KMnO$_4$, MnSO$_4$·H$_2$O, Na$_2$SO$_4$, polyvinylidene fluoride (2%, PVDF), N-methyl-2-pyrrolydone (NMP) were obtained from a Sigma-Aldrich supplier (Taufkirchen, Germany). All reagents were of analytical grade and used as received without further purification. Aqueous solutions were prepared using Milli-Q water with a resistivity of 18.2 MΩ cm^{-1}.

MnO$_2$/C nanocomposite labeled as sample S1 was prepared by the following steps: 2 g of KMnO$_4$ was mixed with 0.1 g of graphite powder and 20 mL of deionized water in an ultrasound bath for 30 min. Then, the reaction mixture was put into a microwave reactor Monowave 300 (Anton Paar, Graz, Austria). The synthesis of MnO$_2$/C was carried out at a temperature of 150 °C for 5 min. After that, the precipitate was filtered out, washed with deionized water, and dried in a vacuum oven at a temperature of 80 °C for 2 h. Equation (1) describes the formation of MnO$_2$ [18]:

$$4KMnO_4 + 3C + H_2O \rightarrow 4MnO_2 + K_2CO_3 + 2KHCO_3 \qquad (1)$$

Another MnO$_2$/C nanocomposite labeled as sample S2 was prepared by the following procedure: at first, the pure MnO$_2$ was prepared. In a typical experiment, 0.063 g of KMnO$_4$

and 0.1 g of MnSO$_4$·H$_2$O was dispersed in 20 mL of deionized water and mixed in an ultrasound bath for 30 min. Then, the reaction mixture was put into a microwave reactor, and the synthesis was carried out at a temperature of 150 °C for 5 min. The precipitate was filtered out, washed with deionized water, and dried in a vacuum oven at 80 °C for two h. Equation (2) shows the formation process of pure MnO$_2$ [58]:

$$2KMnO_4 + 3MnSO_4 + 2H_2O \rightarrow 5MnO_2 + 2H_2SO_4 + K_2SO_4 \qquad (2)$$

Then, 0.01 g of the prepared MnO$_2$ was mixed with 0.1 g of graphite powder and 20 mL of deionized water in an ultrasound bath for 30 min. The synthesis of MnO$_2$/C was carried out under the same conditions as for the sample S1.

The prepared nanocomposites' morphology, structure and composition were characterized using an SEM-focused ion beam facility (Helios Nanolab 650, FEI, Eindhoven, The Netherlands) equipped with an EDX spectrometer (INCA Energy 350 X-Max 20, Oxford Instruments, Oxford, UK). The amount of active material was determined using an ICP optical emission spectrometer Optima700DV (Perkin Elmer, Waltham, MA, USA).

The shape and size of catalyst particles were examined using a Transmission Electron Microscope Tecnai G2 F20 X-TWIN (FEI, Eindhoven, The Netherlands) equipped with an EDAX spectrometer with an r-TEM detector. For microscopic examinations, 10 mg of sample was first sonicated in 1 mL of ethanol for 1 h and then deposited on the Cu grid covered with a continuous carbon film.

XPS measurements were carried out to obtain information about the elemental chemical states and surface composition of powders on the upgraded Vacuum Generator (VG) ESCALAB MKII spectrometer (VG Scientific, UK) fitted with a new XR4 twin anode. The non-monochromatized Al K$_\alpha$ X-ray source was operated at hν = 1486.6 eV with 300 W power (20 mA/15 kV), and the pressure in the analysis chamber was lower than 5×10^{-7} Pa during spectral acquisition. The analyzer work function was determined, assuming the binding energy of the Au4f7/2 peak to be 84.0 eV. The spectra were acquired with an electron analyzer pass energy of 20 eV for narrow scans and resolution of 0.05 eV and with a pass energy of 100 eV for survey spectra. All spectra were recorded at a 90° take-off angle. The spectra calibration, processing, and fitting routines were done using Avantage software (v5.962) provided by Thermo VG Scientific (Waltham, MA, USA). Core level peaks of Mn 2p, Mn 3s, O 1s, and C 1s were recorded and analyzed using a nonlinear Shirley-type background. The calculation of the elemental composition was performed on the basis of Scofield's relative sensitivity factors.

XRD patterns of studied powders were measured using an X-ray diffractometer SmartLab (Rigaku, Japan) equipped with a 9 kW rotating Cu anode X-ray tube. The measurements were performed using Bragg–Brentano geometry with a graphite monochromator on the diffracted beam and a step scan mode with the step size of 0.02° (in 2θ scale) and counting time of 1s per step. The measurements were conducted in the 2θ range 10–75°. Phase identification was performed using software package PDXL (Rigaku, Japan) and ICDD powder diffraction database PDF-4+ (2020 release).

All electrochemical measurements were performed with a three-electrode cell using cyclic voltammetry (CV) and galvanostatic charge/discharge (GCD). The prepared MnO$_2$/C nanocomposites coated on the glassy carbon electrode (GCE) were employed as the working electrode; a Pt sheet as a counter electrode and an Ag/AgCl/KCl electrode were used as a reference. The working electrodes were prepared as follows: the required amount of the active material (MnO$_2$/C) was dispersed ultrasonically in 2% of PVDF in an NMP solution for 1 h. Then, the obtained slurry was pipetted onto the polished surface of GCE and dried in an oven at a temperature of 80 °C for 2 h.

All electrochemical measurements were performed with a Zennium electrochemical workstation (ZAHNER-Elektrik GmbH & Co.KG, Kronach, Germany). Cyclic voltammograms (CVs) were recorded in a 1 M Na$_2$SO$_4$ solution at different scan rates between 10 and 200 mV s^{-1}. All solutions were de-aerated by argon for 15 min before measurements.

The specific capacitance Cs (F g^{-1}) of the electrode material was calculated from the CV test according to the following equation (Equation (3)) [43]:

$$Cs = \frac{1}{m \cdot v \cdot \Delta V} \int i dv, \quad (3)$$

where C_s, is the specific capacitance (F g^{-1}), m—the mass of the active material (g), v—the scan rate of potential (V s^{-1}), ΔV—the range of scan potential (V), and i—the current (A).

Further, galvanostatic charge/discharge cycling was carried out within a potential range between 0 and 1 V at a current density of 1, 2, 5, 10, and 20 A g^{-1}. The C_s was calculated using the following equation (Equation (4)) [55]:

$$Cs = \frac{I \Delta t}{m \cdot \Delta V}, \quad (4)$$

where I is the discharge current (A), Δt is the time for a full discharge (s), and ΔV represents the voltage change during the discharge process (V).

3. Results

The carbon-supported MnO$_2$ nanocomposites were prepared using the microwave-assisted heating method and two different approaches. The MnO$_2$/C nanocomposite, labeled as sample S1, was prepared directly by the microwave-assisted synthesis of mixed KMnO$_4$ and carbon powder components. Another MnO$_2$/C nanocomposite, labeled as sample S2, was prepared in another way: at first, pure MnO$_2$ was obtained by synthesizing KMnO$_4$ and MnSO$_4$. Then, the mixture of the obtained MnO$_2$ and carbon powder was affected by microwave-assisted heating.

SEM images of the prepared MnO$_2$/C samples S1 (a, b) and S2 (c, d) under different magnifications are presented in Figure 1.

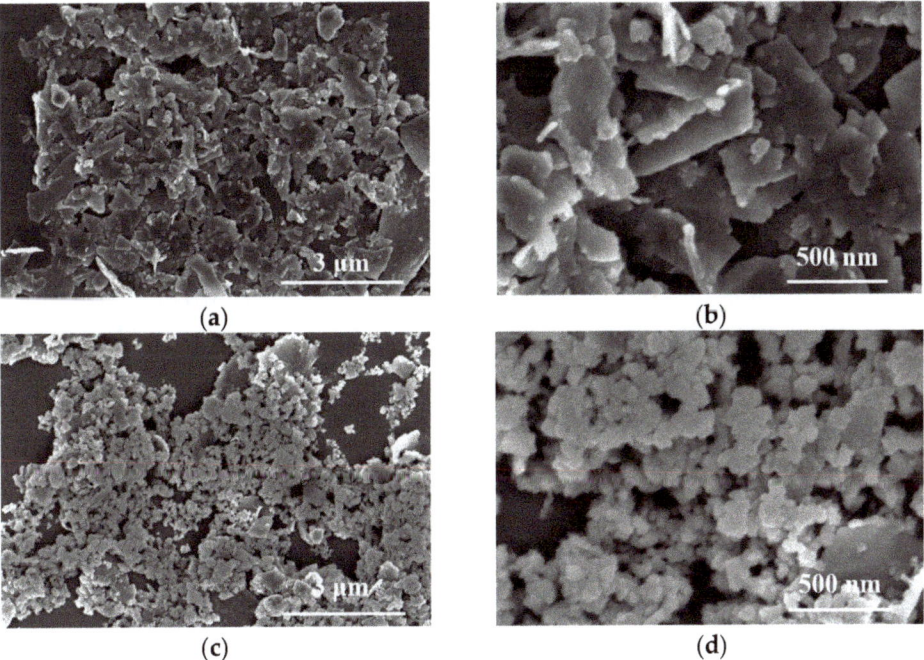

Figure 1. SEM images of MnO$_2$/C: (**a,b**) sample S1; (**c,d**) sample S2 under different magnifications.

As evident, both samples are composed of spherical manganese nanograins located on the carbon surface, but they differ significantly in particle number, size and density. In the case of sample S1, a sparse population of almost separate particles under a low MnO_2 aggregation level is arranged (Figure 1a,b). The size of particles in this sample is close to 20 nm. In the case of sample S2, a nanograins' aggregate forms a large porous network structure with carbon embedded inside (Figure 1c,d). The aggregation level of MnO_2 develops to a large extent without any clear interparticle boundaries.

The samples S1 and S2 were further characterized by TEM analysis. TEM images of the samples S1 and S2 confirm the fibrous morphologies of both samples with the more expressed one for sample S2 (Figure 2). It was found that in the prepared sample S1, the MnO_2 nanoparticles are spherical and are ca. 13–18 nm in size (Figure 2a,b). Furthermore, no large MnO_2 nanoparticles are present within the prepared sample, indicating their negligible aggregation. In the case of sample S2, thin flakelike morphology is observed (Figure 2c,d).

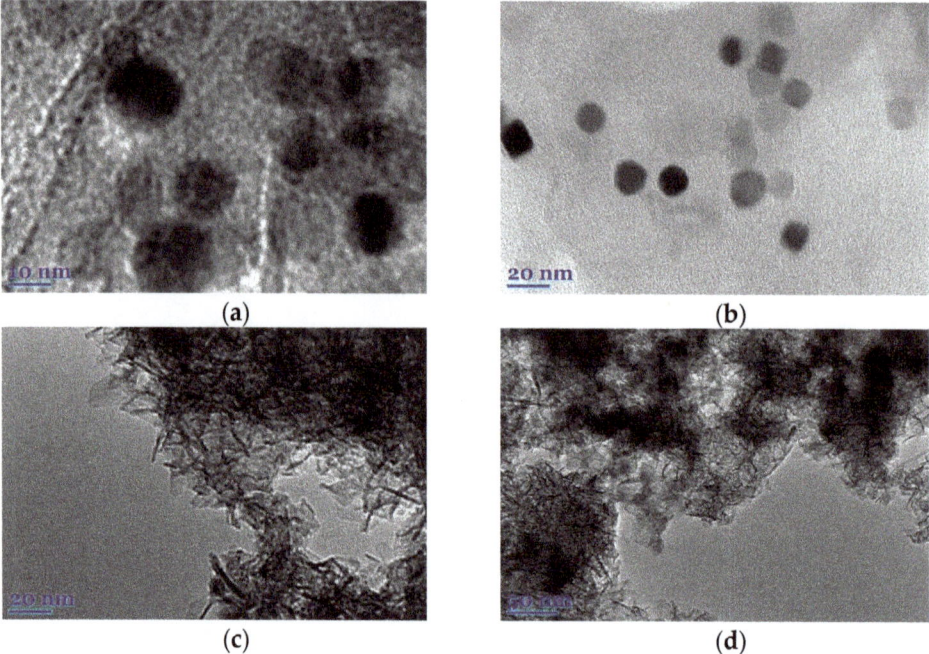

Figure 2. TEM images of MnO_2/C: (**a**,**b**) sample S1; (**c**,**d**) sample S2 under different magnifications.

The chemical composition and the surface electronic state of the prepared MnO_2/C nanocomposites were analyzed using XPS. The C 1s signal and Mn 2p and O 1s peaks were observed in the XPS survey spectra of both samples S1 and S2 (Figure 3a). It indicates the successful synthesis of MnO_2/C, while K content in the samples S1 and S2 was ca. 3.5 and 0.88 at.%, respectively. In both cases, the deconvoluted spectra of Mn show a spin-orbit doublet of the main Mn 2p3/2 and Mn 2p1/2 peaks located at binding energies (E_b) of 642.2 eV and 654.0 eV, respectively, with a spin-energy separation of 11.8 eV (not shown). This value confirms the presence of MnO_2 in the prepared nanocomposites [59–62].

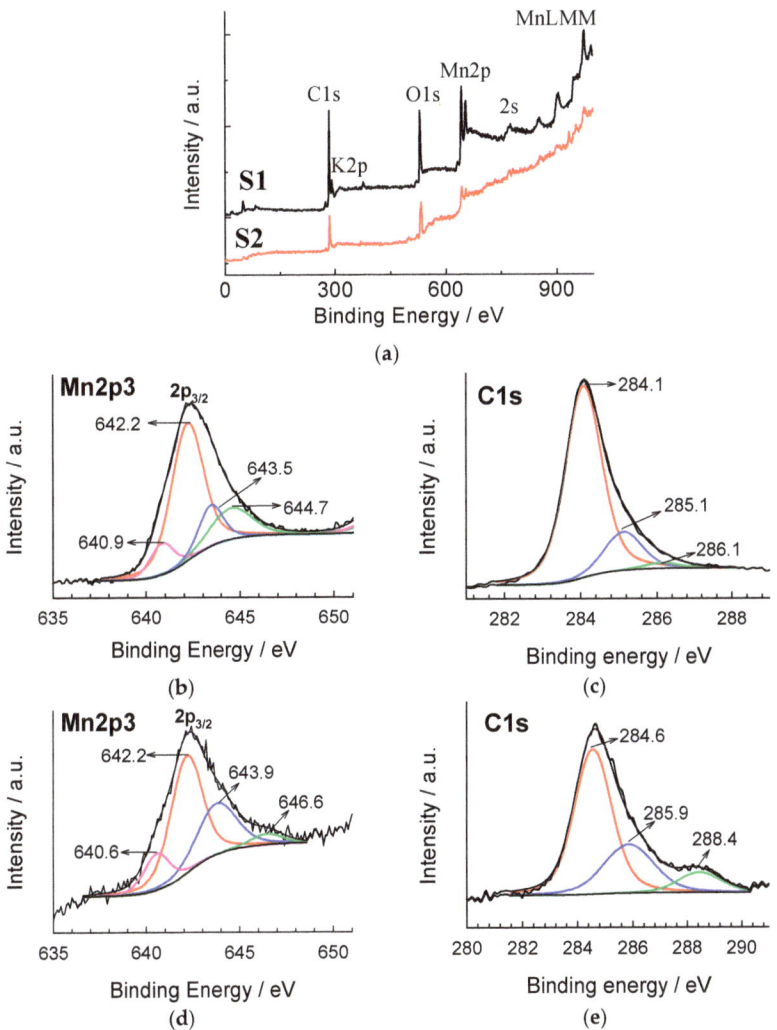

Figure 3. Survey spectra for samples S1 and S2 (**a**). XPS spectra of Mn 2p and C 1s for MnO$_2$/C: (**b**,**c**) sample S1; (**d**,**e**) sample S2.

XPS spectra of Mn2p3/2 and C 1s for MnO$_2$/C samples S1 and S2 are shown in Figure 3b–e. As evident, for both samples, the Mn 2p3/2 peaks were deconvoluted into four peaks at binding energies of 640.6 ± 0.3, 642.2, 643.5, and 644.7 and 646.6 eV, indicating the mixed-valence of manganese oxide phases (Figure 3b,c). Following the data reported in [59–64], the position of deconvoluted Mn 2p3/2 peaks are generally assigned to Mn (IV) or Mn (II) oxidation state at E_b ranging between 641.85–643.0 eV or 640.10–641.12 eV, respectively. Therefore, peaks determined at 640.6 ± 0.3 and 642.2 eV confirm the presence of Mn(II) and Mn(IV) species in the samples S1 and S2 (Figure 3b,c). Moreover, the additional peak at 644.7 eV close to that obtained at 644.9 eV in [65,66] could similarly be assigned to a satellite shake-up peak located at higher E_b values than the main component and is a characteristic feature of the MnO phase Mn 2p core peak maximum at 640.6 ± 0.3 eV [67]. Meanwhile, peaks at 643.5 eV and 644.7 eV based on data in [66] could be related to Mn (VI and VII) species in the samples. There is no Mn 2p3/2 signal

(647 eV) from permanganate ions, suggesting permanganate ions have been reduced to MnO_2 [68]. It should be noted that the dominating fraction in the prepared samples S1 and S2 is the MnO_2 phase and is equal to ~59 and 53%, respectively. At the same time, the MnO (Mn (II)) phase remains significantly lower compared to that determined for the MnO_2 (Mn (IV)) phase.

The high-resolution C 1s spectrum for sample S1 can be deconvoluted into three peaks centered at E_b of 284.1, 285.1 and 286.1 eV (Figure 3d). The first one value could be assigned to carbon atoms C–C; meanwhile, other peaks could be assigned to oxygen functionalized carbon atoms, such as C–O, or C–OH and C=O [60,69].

In the case of the sample S2, the C 1s XPS spectrum could be fitted into three peaks at 284.6, 285.8, and 288.4 eV (Figure 3e), which corresponded to C–C/C=C, C–O, and O–C=O bonds, respectively [70]. The strong peak of C–C/C=C bonds shows that carbon contained high graphitization.

The obtained XRD patterns for both MnO_2/C samples are shown in Figure 4. The presence of broad peaks implied that the synthesized samples S1 and S2 are essentially a mixture of amorphous and nanocrystalline phases. The prominent peaks from both samples (Figure 4) locate at 26°, which can be assigned to the (002) crystal plane of graphitic carbon (ICDD card no. 00-056-0159). The diffraction peaks of α-MnO_2 are indexed according to ICDD card no. 04-005-4884, indicating a tetragonal unit cell with lattice parameters of a = b = 9.82 Å and c = 2.85 Å. The synthesized powders are composed of small crystallites with an average size of about 3.2 ± 0.3 nm.

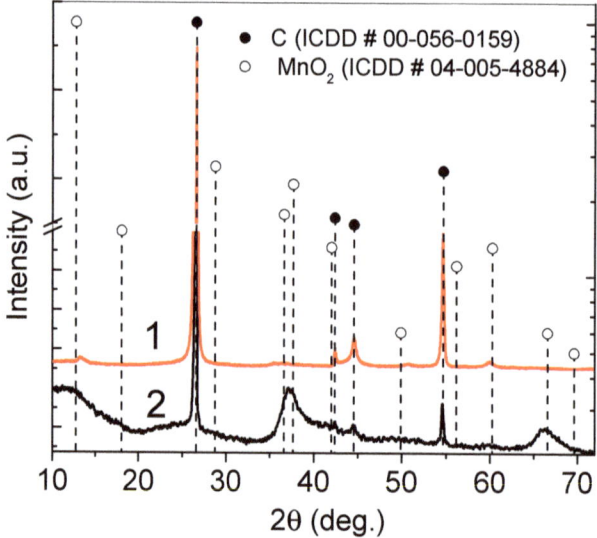

Figure 4. XRD patterns of samples S1 (1 pattern) and sample S2 (2 pattern).

The electrochemical performance of samples S1 and S2 was evaluated from the cyclic voltammetry and galvanostatic charge/discharge measurements using a three-electrode system in a 1 M Na_2SO_4 solution. Figure 5 shows the CV curves of the sample S1 (a), sample S2 (b), and pure carbon (c) at the scan rates of 10, 50, 100, and 200 mV s^{-1}. No obvious peaks are observed in all the CV curves. This indicates that the electrodes are charged and discharge at a constant rate over the complete cycle.

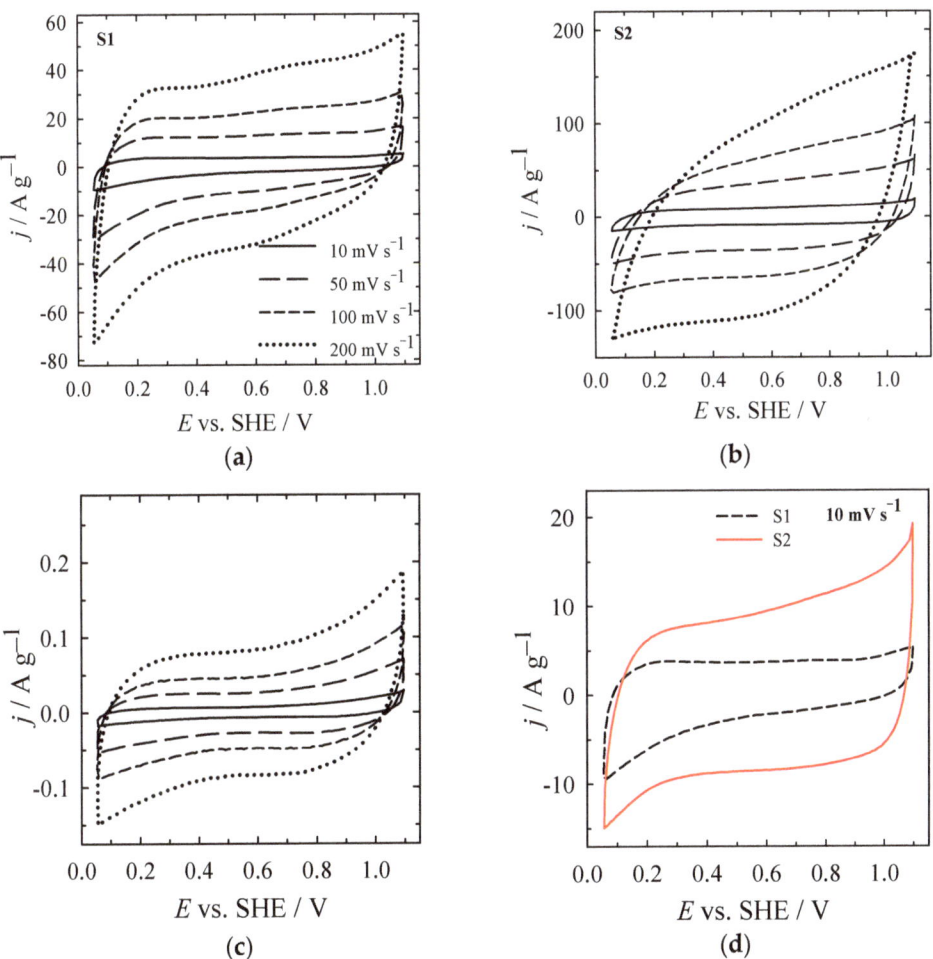

Figure 5. CVs of: (**a**) sample S1; (**b**) sample S2; (**c**) carbon were recorded in 1 M Na_2SO_4 at different scan rates. (**d**) CVs of samples S1 and S2 at 10 mV s^{-1}.

Both sample S1 and carbon show symmetrical rectangular shapes, which indicates the ideal capacitive behavior of those samples (Figure 5a,c). In the case of sample S2, deviations in the rectangularity of CV curves occur (Figure 5b). It can be seen that the current response of sample S2 is significantly higher as compared with that of sample S1 and carbon (Figure 5a–c). It is clearly seen that the sample S2 shows a significantly higher capacitive behavior as compared with that of sample S1 (Figure 5d).

The calculated C_s values for the sample S2 were 980.7, 743.2, 641.0, and 536.6 F g^{-1} at scan rates of 10, 50, 100, and 200 mV s^{-1}, respectively (Figure 6). Meanwhile, for the sample S1, Cs values were 535.8, 349.6, 275.2, and 209.7 F g^{-1} at scan rates of 10, 50, 100, and 200 mV s^{-1}. Those values were found to be ca. 1.8–2.6 times lower than those obtained for sample S2.

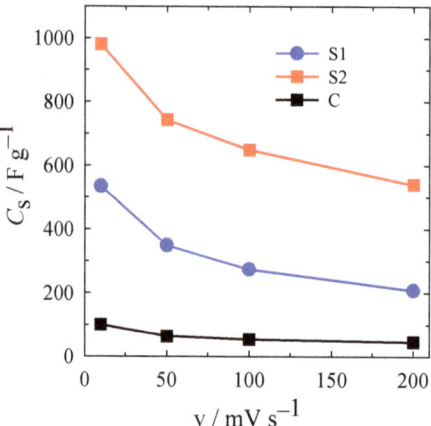

Figure 6. Specific capacitances of the sample S1, sample S2, and carbon obtained from CV curves.

Comparisons of the supercapacitive behavior of various MnO_2-based electrode materials reported in the literature and the present work are listed in Table 1, exhibiting the high specific capacitance of our prepared electrode materials.

Table 1. Comparisons of specific capacitance for various MnO_2-based electrode materials.

Materials	Scan Rate, $mV\ s^{-1}$	Specific Capacitance, $F\ g^{-1}$	Ref.
MnO_2/C (S2)	10	980.7	This work
MnO_2/C (S1)	10	535.8	This work
Self-branched α-MnO_2/δ-MnO_2 heterojunction nanowires	10	152.0	[22]
MnO_2	5	380.0	[24]
MnO_2	10	154.0	[29]
MnO_2/3D-PC	1	416.0	[47]
MnO_2	5	547.0	[68]
Ultra-long MnO_2 nanowires	2	495.0	[69]
MnO_2 NPs/Ni foam	5	549.0	[71]
MnO_2/MWCNT	2	553.0	[72]

Galvanostatic charge/discharge curves for the sample S2 measured at different current densities of 1, 2, 5, 10, and 20 A g^{-1} are shown in Figure 7a. The shapes of the curves show a typical triangular symmetrical distribution with a slight curvature. This result indicates a combination of electric double-layer and pseudocapacitive contributions. Specific capacitance values were calculated from the discharge test. It was found that the sample S2 can deliver high Cs values of 949.3, 719.3, 480.8, 406.7, and 371.5 F g^{-1} at a current density of 1, 2, 5, 10, and 20 A g^{-1} (Figure 7b). The long-term stability of the charge/discharge process was also performed on this sample S2 at a high current density of 20 A g^{-1} up to 100 cycles (Figure 7c). As evident, this electrode showed excellent long-term stability with 93% retention of its initial capacitance value during 100 cycles.

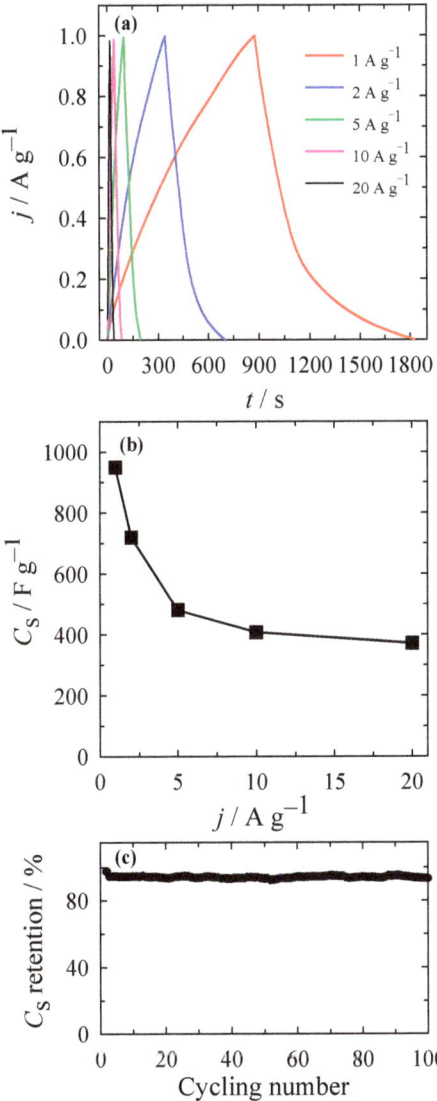

Figure 7. (a) Galvanostatic charge/discharge curves of the sample S2 measured at different constant current densities of 1–20 A g^{-1}; (b) Specific capacitance obtained from galvanostatic charge/discharge curves with different current densities.

4. Conclusions

We have successfully fabricated carbon-supported MnO$_2$ nanocomposites via a simple microwave-assisted heating method. Different architecture containing MnO$_2$ nanocomposites demonstrates improved conductivity, which is a key limitation in pseudocapacitors. The electrochemical measurements revealed that (due to this conductivity) MnO$_2$/C nanocomposites, especially those prepared via a two-step procedure, exhibit excellent electrochemical performance, including a high specific capacitance of 980.7 F g^{-1}. Moreover, the specific capacitance retention was 93% after 100 cycles at 20 A g^{-1}, indicating good electrochemical stability. The obtained results demonstrate that the prepared MnO$_2$/C nanocomposites should be a promising electrode material for supercapacitor applications and could be further extended to fabricate other materials for supercapacitors.

Author Contributions: This study was conducted through the contributions of all authors. Conceptualization, J.J.; L.T.-T. and E.N.; methodology, G.S., A.D., V.P. and V.J.; investigation, J.V., G.S. and A.D.; writing—original draft preparation, J.J. and D.S.; writing—review and editing, L.T.-T. and E.N.; visualization, D.S., J.V., V.P. and V.J. All authors have read and agreed to the published version of the manuscript.

Funding: This research was funded by the European Social Fund under the No 09.3.3-LMT-K-712-02-0142 "Development of Competencies of Scientists, other Researchers, and Students through Practical Research Activities" measure.

Institutional Review Board Statement: Not applicable.

Informed Consent Statement: Not applicable.

Conflicts of Interest: The authors declare no conflict of interest.

References

1. Gopi, C.V.M.; Vinodh, R.; Sambasivam, S.; Obaidat, I.M.; Kim, H.-J. Recent progress of advanced energy storage materials for flexible and wearable supercapacitor: From design and development to applications. *J. Energy Storage* **2020**, *27*, 101035. [CrossRef]
2. Poonam; Sharma, K.; Arora, A.; Tripathi, S.K. Review of supercapacitors: Materials and devices. *J. Energy Storage* **2019**, *21*, 801–825. [CrossRef]
3. Liu, C.; Li, F.; Ma, L.-P.; Cheng, H.-M. Advanced Materials for Energy Storage. *Adv. Mater.* **2010**, *22*, E28–E62. [CrossRef] [PubMed]
4. Miller, E.E.; Hua, Y.; Tezel, F.H. Materials for energy storage: Review of electrode materials and methods of increasing capacitance for supercapacitors. *J. Energy Storage* **2018**, *20*, 30–40. [CrossRef]
5. Wang, F.; Wu, X.; Yuan, X.; Liu, Z.; Zhang, Y.; Fu, L.; Zhu, Y.; Zhou, Q.; Wu, Y.; Huang, W. Latest advances in supercapacitors: From new electrode materials to novel device designs. *Chem. Soc. Rev.* **2017**, *46*, 6816–6854. [CrossRef] [PubMed]
6. Liu, C.; Yan, X.; Hu, F.; Gao, G.; Wu, G.; Yang, X. Toward Superior Capacitive Energy Storage: Recent Advances in Pore Engineering for Dense Electrodes. *Adv. Mater.* **2018**, *30*, e1705713. [CrossRef] [PubMed]
7. Simon, P.; Gogotsi, Y. Materials for electrochemical capacitors. *Nat. Mater.* **2008**, *7*, 845–854. [CrossRef]
8. Guo, W.; Yu, C.; Li, S.; Wang, Z.; Yu, J.; Huang, H.; Qiu, J. Strategies and insights towards the intrinsic capacitive properties of MnO_2 for supercapacitors: Challenges and perspectives. *Nano Energy* **2019**, *57*, 459–472. [CrossRef]
9. Wang, G.; Zhang, L.; Zhang, J. A review of electrode materials for electrochemical supercapacitors. *Chem. Soc. Rev.* **2012**, *41*, 797–828. [CrossRef]
10. Muzaffar, A.; Ahamed, M.B.; Deshmukh, K.; Thirumalai, J. A review on recent advances in hybrid supercapacitors: Design, fabrication and applications. *Renew. Sustain. Energy Rev.* **2019**, *101*, 123–145. [CrossRef]
11. Afif, A.; Rahman, S.M.; Azad, A.T.; Zaini, J.; Islam, A.; Azad, A. Advanced materials and technologies for hybrid supercapacitors for energy storage—A review. *J. Energy Storage* **2019**, *25*, 100852. [CrossRef]
12. Xu, W.; Jiang, Z.; Yang, Q.; Huo, W.; Javed, M.S.; Li, Y.; Huang, L.; Gu, X.; Hu, C. Approaching the lithium-manganese oxides' energy storage limit with Li_2MnO_3 nanorods for high-performance supercapacitor. *Nano Energy* **2018**, *43*, 168–176. [CrossRef]
13. Xia, H.; Hong, C.; Li, B.; Zhao, B.; Lin, Z.; Zheng, M.; Savilov, S.V.; Aldoshin, S.M. Facile Synthesis of Hematite Quantum-Dot/Functionalized Graphene-Sheet Composites as Advanced Anode Materials for Asymmetric Supercapacitors. *Adv. Funct. Mater.* **2015**, *25*, 627–635. [CrossRef]
14. Qiu, T.; Luo, B.; Giersig, M.; Akinoglu, E.M.; Hao, L.; Wang, X.; Shi, L.; Jin, M.; Zhi, L. Au@MnO_2 Core-Shell Nanomesh Electrodes for Transparent Flexible Supercapacitors. *Small* **2014**, *10*, 4136–4141. [CrossRef]
15. Huang, M.; Zhang, Y.; Li, F.; Zhang, L.; Wen, Z.; Liu, Q. Facile synthesis of hierarchical Co_3O_4@MnO_2 core–shell arrays on Ni foam for asymmetric supercapacitors. *J. Power Sources* **2014**, *252*, 98–106. [CrossRef]
16. Radhamani, A.V.; Shareef, K.M.; Rao, M.S.R. ZnO@MnO_2 Core-Shell Nanofiber Cathodes for High Performance Asymmetric Supercapacitors. *ACS Appl. Mater. Interfaces* **2016**, *8*, 30531–30542. [CrossRef]
17. Wang, Y.; Zeng, J.; Li, J.; Cui, X.; Al-Enizi, A.M.; Zhang, L.; Zheng, G. One-dimensional nanostructures for flexible supercapacitors. *J. Mater. Chem. A* **2015**, *3*, 16382–16392. [CrossRef]
18. Liu, L.; Niu, Z.; Chen, J. Unconventional supercapacitors from nanocarbon-based electrode materials to device configurations. *Chem. Soc. Rev.* **2016**, *45*, 4340–4363. [CrossRef]
19. Huang, M.; Li, F.; Dong, F.; Zhang, Y.X.; Zhang, L. MnO_2-based nanostructures for high-performance supercapacitors. *J. Mater. Chem. A* **2015**, *3*, 21380–21423. [CrossRef]
20. Huang, M.; Zhang, Y.; Li, F.; Zhang, L.; Ruoff, R.S.; Wen, Z.; Liu, Q. Self-Assembly of Mesoporous Nanotubes Assembled from Interwoven Ultrathin Birnessite-type MnO_2 Nanosheets for Asymmetric Supercapacitors. *Sci. Rep.* **2015**, *4*, 3878. [CrossRef]

21. Yin, B.; Zhang, S.; Jiao, Y.; Liu, Y.; Qu, F.; Wu, X. Facile synthesis of ultralong MnO_2 nanowires as high performance supercapacitor electrodes and photocatalysts with enhanced photocatalytic activities. *CrystEngComm* **2014**, *16*, 9999–10005. [CrossRef]
22. Zhu, C.; Yang, L.; Seo, J.K.; Zhang, X.; Wang, S.; Shin, J.; Chao, D.; Zhang, H.; Meng, Y.S.; Fan, H.J. Self-branched α-MnO_2/δ-MnO_2 heterojunction nanowires with enhanced pseudocapacitance. *Mater. Horiz.* **2017**, *4*, 415–422. [CrossRef]
23. Lv, Q.; Sun, H.; Li, X.; Xiao, J.; Xiao, F.; Liu, L.; Luo, J.; Wang, S. Ultrahigh capacitive performance of three-dimensional electrode nanomaterials based on α-MnO_2 nanocrystallines induced by doping Au through Å-scale channels. *Nano Energy* **2016**, *21*, 39–50. [CrossRef]
24. Kang, J.; Hirata, A.; Kang, L.; Zhang, X.; Hou, Y.; Chen, L.; Li, C.; Fujita, T.; Akagi, K.; Chen, M. Enhanced Supercapacitor Performance of MnO_2 by Atomic Doping. *Angew. Chem. Int. Ed.* **2013**, *52*, 1664–1667. [CrossRef] [PubMed]
25. Hu, Z.; Xiao, X.; Chen, C.; Li, T.; Huang, L.; Zhang, C.; Su, J.; Miao, L.; Jiang, J.; Zhang, Y.; et al. Al-doped α-MnO_2 for high mass-loading pseudocapacitor with excellent cycling stability. *Nano Energy* **2015**, *11*, 226–234. [CrossRef]
26. Su, X.; Yu, L.; Cheng, G.; Zhang, H.; Sun, M.; Zhang, L.; Zhang, J. Controllable hydrothermal synthesis of Cu-doped δ-MnO_2 films with different morphologies for energy storage and conversion using supercapacitors. *Appl. Energy* **2014**, *134*, 439–445. [CrossRef]
27. Peng, R.; Wu, N.; Zheng, Y.; Huang, Y.; Luo, Y.; Yu, P.; Zhuang, L. Large-Scale Synthesis of Metal-Ion-Doped Manganese Dioxide for Enhanced Electrochemical Performance. *ACS Appl. Mater. Interfaces* **2016**, *8*, 8474–8480. [CrossRef]
28. Wang, Z.; Wang, F.; Li, Y.; Hu, J.; Lu, Y.; Xu, M. Interlinked multiphase Fe-doped MnO_2 nanostructures: A novel design for enhanced pseudocapacitive performance. *Nanoscale* **2016**, *8*, 7309–7317. [CrossRef]
29. Tang, C.-L.; Wei, X.; Jiang, Y.-M.; Wu, X.-Y.; Wang, K.-X.; Chen, J.-S.; Han, L.-N. Cobalt-Doped MnO_2 Hierarchical Yolk–Shell Spheres with Improved Supercapacitive Performance. *J. Phys. Chem. C* **2015**, *119*, 8465–8471. [CrossRef]
30. Hashem, A.M.A.; Abuzeid, H.M.; Narayanan, N.; Ehrenberg, H.; Julien, C. Synthesis, structure, magnetic, electrical and electrochemical properties of Al, Cu and Mg doped MnO_2. *Mater. Chem. Phys.* **2011**, *130*, 33–38. [CrossRef]
31. Zhang, X.; Meng, X.; Gong, S.; Li, P.; Jin, L.; Cao, Q. Synthesis and characterization of 3D MnO_2/carbon microtube bundle for supercapacitor electrodes. *Mater. Lett.* **2016**, *179*, 73–77. [CrossRef]
32. Pandolfo, A.; Hollenkamp, A. Carbon properties and their role in supercapacitors. *J. Power Sources* **2006**, *157*, 11–27. [CrossRef]
33. Frackowiak, E. Carbon materials for supercapacitor application. *Phys. Chem. Chem. Phys.* **2007**, *9*, 1774–1785. [CrossRef] [PubMed]
34. Zhang, L.; Zhao, X.S. Carbon-based materials as supercapacitor electrodes. *Chem. Soc. Rev.* **2009**, *38*, 2520–2531. [CrossRef] [PubMed]
35. Vattikuti, S.P.; Reddy, B.P.; Byon, C.; Shim, J. Carbon/CuO nanosphere-anchored g-C_3N_4 nanosheets as ternary electrode material for supercapacitors. *J. Solid State Chem.* **2018**, *262*, 106–111. [CrossRef]
36. Vattikuti, S.V.P.; Police, A.K.R.; Shim, J.; Byon, C. Sacrificial-template-free synthesis of core-shell C@Bi_2S_3 heterostructures for efficient supercapacitor and H_2 production applications. *Sci. Rep.* **2018**, *8*, 4194. [CrossRef]
37. Vattikuti, S.P.; Devarayapalli, K.C.; Dang, N.N.; Shim, J. 1D/1D $Na_2Ti_3O_7$/SWCNTs electrode for split-cell-type asymmetric supercapacitor device. *Ceram. Int.* **2021**, *47*, 11602–11610. [CrossRef]
38. Ping, Y.; Liu, Z.; Li, J.; Han, J.; Yang, Y.; Xiong, B.; Fang, P.; He, C. Boosting the performance of supercapacitors based hierarchically porous carbon from natural Juncus effuses by incorporation of MnO_2. *J. Alloys Compd.* **2019**, *805*, 822–830. [CrossRef]
39. Feng, D.-Y.; Sun, Z.; Huang, Z.-H.; Cai, X.; Song, Y.; Liu, X.-X. Highly loaded manganese oxide with high rate capability for capacitive applications. *J. Power Sources* **2018**, *396*, 238–245. [CrossRef]
40. Sun, L.; Li, N.; Zhang, S.; Yu, X.; Liu, C.; Zhou, Y.; Han, S.; Wang, W.; Wang, Z. Nitrogen-containing porous carbon/α-MnO_2 nanowires composite electrode towards supercapacitor applications. *J. Alloys Compd.* **2019**, *789*, 910–918. [CrossRef]
41. Su, X.; Yu, L.; Cheng, G.; Zhang, H.; Sun, M.; Zhang, X. High-performance α-MnO_2 nanowire electrode for supercapacitors. *Appl. Energy* **2015**, *153*, 94–100. [CrossRef]
42. Lei, R.; Zhang, H.; Lei, W.; Li, D.; Fang, Q.; Ni, H.; Gu, H. MnO_2 nanowires electrodeposited on freestanding graphenated carbon nanotubes as binder-free electrodes with enhanced supercapacitor performance. *Mater. Lett.* **2019**, *249*, 140–142. [CrossRef]
43. Meng, X.; Lu, L.; Sun, C. Green Synthesis of Three-Dimensional MnO_2/Graphene Hydrogel Composites as a High-Performance Electrode Material for Supercapacitors. *ACS Appl. Mater. Interfaces* **2018**, *10*, 16474–16481. [CrossRef] [PubMed]
44. Qiu, Y.; Xu, P.; Guo, B.; Cheng, Z.; Fan, H.; Yang, M.; Yang, X.; Li, J. Electrodeposition of manganese dioxide film on activated carbon paper and its application in supercapacitors with high rate capability. *RSC Adv.* **2014**, *4*, 64187–64192. [CrossRef]
45. Cheng, H.; Zhao, S.; Yi, F.; Gao, A.; Shu, D.; Ao, Z.; Huang, S.; Zhou, X.; He, C.; Li, S.; et al. Supramolecule-assisted synthesis of in-situ carbon-coated MnO_2 nanosphere for supercapacitors. *J. Alloys Compd.* **2019**, *779*, 550–556. [CrossRef]
46. Yang, M.; Kim, D.S.; Hong, S.B.; Sim, J.-W; Kim, J.; Kim, S.-S.; Choi, B.G. MnO_2 Nanowire/Biomass-Derived Carbon from Hemp Stem for High-Performance Supercapacitors. *Langmuir* **2017**, *33*, 5140–5147. [CrossRef]
47. Wang, L.; Zheng, Y.; Chen, S.; Ye, Y.; Xu, F.; Tan, H.; Li, Z.; Hou, H.; Song, Y. Three-Dimensional Kenaf Stem-Derived Porous Carbon/MnO_2 for High-Performance Supercapacitors. *Electrochim. Acta* **2014**, *135*, 380–387. [CrossRef]
48. Wang, H.; Fu, Q.; Pan, C. Green mass synthesis of graphene oxide and its MnO_2 composite for high performance supercapacitor. *Electrochim. Acta* **2019**, *312*, 11–21. [CrossRef]
49. Nie, G.; Lu, X.; Chi, M.; Gao, M.; Wang, C. General synthesis of hierarchical C/MO_x@MnO_2 (M = Mn, Cu, Co) composite nanofibers for high-performance supercapacitor electrodes. *J. Colloid Interface Sci.* **2018**, *509*, 235–244. [CrossRef]

50. Wang, X.; Chen, S.; Li, D.; Sun, S.; Peng, Z.; Komarneni, S.; Yang, D. Direct Interfacial Growth of MnO_2 Nanostructure on Hierarchically Porous Carbon for High-Performance Asymmetric Supercapacitors. *ACS Sustain. Chem. Eng.* **2018**, *6*, 633–641. [CrossRef]
51. Kang, H.G.; Jeong, J.; Hong, S.B.; Lee, G.Y.; Kim, D.H.; Kim, J.W.; Choi, B.G. Scalable exfoliation and activation of graphite into porous graphene using microwaves for high–performance supercapacitors. *J. Alloys Compd.* **2019**, *770*, 458–465. [CrossRef]
52. Bi, Y.; Nautiyal, A.; Zhang, H.; Yan, H.; Luo, J.; Zhang, X. Facile and ultrafast solid-state microwave approach to MnO_2-NW@Graphite nanocomposites for supercapacitors. *Ceram. Int.* **2018**, *44*, 5402–5410. [CrossRef]
53. Wang, F.; Zhou, Q.; Li, G.; Wang, Q. Microwave preparation of 3D flower-like MnO_2/$Ni(OH)_2$/nickel foam composite for high-performance supercapacitors. *J. Alloys Compd.* **2017**, *700*, 185–190. [CrossRef]
54. Zhang, X.; Miao, W.; Li, C.; Sun, X.; Wang, K.; Ma, Y. Microwave-assisted rapid synthesis of birnessite-type MnO_2 nanoparticles for high performance supercapacitor applications. *Mater. Res. Bull.* **2015**, *71*, 111–115. [CrossRef]
55. Zhang, X.; Sun, X.; Zhang, H.; Zhang, D.; Ma, Y. Microwave-assisted reflux rapid synthesis of MnO_2 nanostructures and their application in supercapacitors. *Electrochim. Acta* **2013**, *87*, 637–644. [CrossRef]
56. Meher, S.K.; Rao, G.R. Enhanced activity of microwave synthesized hierarchical MnO_2 for high performance supercapacitor applications. *J. Power Sources* **2012**, *215*, 317–328. [CrossRef]
57. Li, Y.; Wang, J.; Zhang, Y.; Banis, M.N.; Liu, J.; Geng, D.; Li, R.; Sun, X. Facile controlled synthesis and growth mechanisms of flower-like and tubular MnO_2 nanostructures by microwave-assisted hydrothermal method. *J. Colloid Interface Sci.* **2012**, *369*, 123–128. [CrossRef]
58. Bagotsky, V.S.; Skundin, A.M.; Volfkovich, Y.M. *Electrochemical Power Sources—Batteries, Fuel Cells, and Supercapacitors*; John Wiley & Sons: Hoboken, NJ, USA, 2015.
59. Audi, A.A.; Sherwood, P. Valence-band X-ray photoelectron spectroscopic studies of manganese and its oxides interpreted by cluster and band structure calculations. *Surf. Interface Anal.* **2002**, *33*, 274–282. [CrossRef]
60. Zhou, D.; Lin, H.; Zhang, F.; Niu, H.; Cui, L.; Wang, Q.; Qu, F. Freestanding MnO_2 nanoflakes/porous carbon nanofibers for high-performance flexible supercapacitor electrodes. *Electrochim. Acta* **2015**, *161*, 427–435. [CrossRef]
61. Ramírez, A.; Hillebrand, P.; Stellmach, D.; May, M.; Bogdanoff, P.; Fiechter, S. Evaluation of MnO_x, Mn_2O_3, and Mn_3O_4 Electrodeposited Films for the Oxygen Evolution Reaction of Water. *J. Phys. Chem. C* **2014**, *118*, 14073–14081. [CrossRef]
62. NIST X-ray Photoelectron Spectroscopy Database. Available online: https://srdata.nist.gov/xps/ (accessed on 11 May 2021).
63. Ilton, E.S.; Post, J.E.; Heaney, P.J.; Ling, F.T.; Kerisit, S.N. XPS determination of Mn oxidation states in Mn (hydr)oxides. *Appl. Surf. Sci.* **2016**, *366*, 475–485. [CrossRef]
64. Baer, D.R.; Artyushkova, K.; Cohen, H.; Easton, C.D.; Engelhard, M.; Gengenbach, T.R.; Greczynski, G.; Mack, P.; Morgan, D.J.; Roberts, A. XPS guide: Charge neutralization and binding energy referencing for insulating samples. *J. Vac. Sci. Technol. A* **2020**, *38*, 031204. [CrossRef]
65. Beyazay, T.; Oztuna, F.E.S.; Unal, U. Self-Standing Reduced Graphene Oxide Papers Electrodeposited with Manganese Oxide Nanostructures as Electrodes for Electrochemical Capacitors. *Electrochim. Acta* **2019**, *296*, 916–924. [CrossRef]
66. Biesinger, M.C.; Payne, B.P.; Grosvenor, A.P.; Lau, L.W.; Gerson, A.R.; Smart, R.S. Resolving surface chemical states in XPS analysis of first row transition metals, oxides and hydroxides: Cr, Mn, Fe, Co and Ni. *Appl. Surf. Sci.* **2011**, *257*, 2717–2730. [CrossRef]
67. Grissa, R.; Martinez, H.; Cotte, S.; Galipaud, J.; Pecquenard, B.; Le Cras, F. Thorough XPS analyses on overlithiated manganese spinel cycled around the 3V plateau. *Appl. Surf. Sci.* **2017**, *411*, 449–456. [CrossRef]
68. Dong, X.; Shen, W.; Gu, J.; Xiong, L.; Zhu, Y.; Li, A.H.; Shi, J. MnO_2-Embedded-in-Mesoporous-Carbon-Wall Structure for Use as Electrochemical Capacitors. *J. Phys. Chem. B* **2006**, *110*, 6015–6019. [CrossRef]
69. Singu, B.S.; Hong, S.E.; Yoon, K.R. Ultra-thin and ultra-long α-MnO_2 nanowires for pseudocapacitor material. *J. Solid State Electrochem.* **2017**, *21*, 3215–3220. [CrossRef]
70. Yang, Y.; Niu, H.; Qin, F.; Guo, Z.; Wang, J.; Ni, G.; Zuo, P.; Qu, S.; Shen, W. MnO_2 doped carbon nanosheets prepared from coal tar pitch for advanced asymmetric supercapacitor. *Electrochim. Acta* **2020**, *354*, 136667. [CrossRef]
71. Edison, T.N.J.I.; Atchudan, R.; Karthik, N.; Xiong, D.; Lee, Y.R. Direct electro-synthesis of MnO_2 nanoparticles over nickel foam from spent alkaline battery cathode and its supercapacitor performance. *J. Taiwan Inst. Chem. Eng.* **2019**, *97*, 414–423. [CrossRef]
72. Xue, C.; Hao, Y.; Luan, Q.; Wang, E.; Ma, X.; Hao, X. Porous manganese dioxide film built from arborization-like nanoclusters and its superior electrochemical supercapacitance with attractive cyclic stability. *Electrochim. Acta* **2019**, *296*, 94–101. [CrossRef]

Article

Effect of Poly(Titanium Oxide) on the Viscoelastic and Thermophysical Properties of Interpenetrating Polymer Networks

Tamara Tsebriienko [1] and Anatoli I. Popov [2],*

[1] Institute of Physics, National Academy of Sciences of Ukraine, 46, pr. Nauky, 03028 Kyiv, Ukraine; mara8@ukr.net

[2] Institute of Solid State Physics, University of Latvia, Kengaraga 8, LV-1063 Riga, Latvia

* Correspondence: popov@latnet.com

Abstract: The influence of poly(titanium oxide) obtained using the sol-gel method in 2-hydroxyethyl methacrylate medium on the viscoelastic and thermophysical properties of interpenetrating polymer networks (IPNs) based on cross-linked polyurethane (PU) and poly(hydroxyethyl methacrylate) (PHEMA) was studied. It was found that both the initial (IPNs) and organo-inorganic interpenetrating polymer networks (OI IPNs) have a two-phase structure by using methods of dynamic mechanical analysis (DMA) and differential scanning calorimetry (DSC). The differential scanning calorimetry methods and scanning electron microscopy (SEM) showed that the presence of poly(titanium oxide) increases the compatibility of the components of IPNs. It was found that an increase in poly(titanium oxide) content leads to a decrease in the intensity of the relaxation maximum for PHEMA phase and an increase in the effective crosslinking density due to the partial grafting of the inorganic component to acrylate. It was shown that the topology of poly(titanium oxide) structure has a significant effect on the relaxation behavior of OI IPNs samples. According to SEM, a uniform distribution of the inorganic component in the polymer matrix is observed without significant aggregation.

Keywords: poly(titanium oxide); sol-gel method; interpenetrating polymer networks; 2-hydroxyethyl methacrylate; polyurethane

Citation: Tsebriienko, T.; Popov, A.I. Effect of Poly(Titanium Oxide) on the Viscoelastic and Thermophysical Properties of Interpenetrating Polymer Networks. *Crystals* **2021**, *11*, 794. https://doi.org/10.3390/cryst11070794

Academic Editors: Kil Sik Min and Ulli Englert

Received: 28 February 2021
Accepted: 4 July 2021
Published: 7 July 2021

Publisher's Note: MDPI stays neutral with regard to jurisdictional claims in published maps and institutional affiliations.

Copyright: © 2021 by the authors. Licensee MDPI, Basel, Switzerland. This article is an open access article distributed under the terms and conditions of the Creative Commons Attribution (CC BY) license (https://creativecommons.org/licenses/by/4.0/).

1. Introduction

In the last decade, due to the high rates of development of various industries, there is a need to obtain materials with the desired set of new operational properties. Therefore, the development of new hybrid organic-inorganic nanocomposites is a topical area of polymer chemistry [1–12]. In this case, the concept of "hybrid" emphasizes the chemical nature of the interaction of system components [13]. Such materials demonstrate not only improved properties of the organic matrix, but also the appearance of new, specific properties that are not inherent to the organic component due to the presence of an inorganic one. Particular attention is drawn to organic-inorganic materials containing titanium, due to their potential application as photocatalyst and membrane, as well as in solar and fuel cells, biomedicine and in other areas, where their unique optical properties can be used [14–30]. In particular, in [25], one of the most extensively studied polymers in biomedical applications, namely, PHEMA incorporated with TiO_2 nanoparticles and the appropriate bioactive behavior of such nanocomposite was investigated in detail. It is especially important to mention here that the formation of bone-like apatite, which is a necessary condition for a synthetic material to be considered bioactive [25]. The optically transparent TiO_2 particles incorporated in PHEMA thin films were prepared in [26], where their photocatalytic activity was successfully demonstrated. Another important result that needs to be mentioned is the synthesis of PHEMA hydrogels containing low concentration of TiO_2 nanoparticles, which show their potential use for cell implantation experiments in vivo [27]. The enhanced mechanical and thermal properties of TiO_2 nanocomposites reinforced with photo-resin via 3-dimensional printing were recently shown in [28]. Another interesting report [29] is about the excellent UV shielding properties of poly(methylmethacrylate) PMMA/oxide nanocomposites with different types of nanoparticles, which could be used in variety of applications such as sunscreens, aerospace, and several other fields related to UV photodegradation. It is important to note that the thermal and mechanical properties of PMMA-titania nanocomposites and their degradation behavior was discussed in detail in [30]. Note also that nanoporous anatase layers may become interesting for nonlinear optical applications [31].

It is also known from the literature that poly(titanium oxide) gels obtained using the sol-gel method have high photocatalytic activity [32]. They exhibit unique optical properties, i.e., UV-induced transition $Ti^{4+} \leftrightarrow Ti^{3+}$. When light is absorbed in a semiconductor, electrons in the conduction band (CB) and positive holes in the valence band (VB) are formed. Light holes can rapidly move to the metal oxide network interface and initiate chemical reactions with environmental molecules. Heavier CB-electrons move on a shorter distance and then are trapped on Ti^{4+} of the metal oxide network as Ti^{3+}, which can also be situated at the interface. These trapped electrons can be visualized with a strong absorption in the UV-visible-nearIR spectral range with a maximum at $\sim 600^{+200}_{-100}$ nm, which depends on the sample preparation conditions. The disadvantage of gel materials that limits their use is their mechanical instability, and the advantages for applications in photonics are their transparency and unique photosensitivity, therefore it is actually the choice of the polymer matrix to stabilize the gel while maintaining its structure and, accordingly, the sensitivity. The problem can be solved by creating the hybrid optically transparent organic-inorganic material, which combines valuable properties of organic and inorganic components.

In light of this problem, the synthesis of TiO_2-containing organic-inorganic copolymers seems promising. In the previous research, the materials containing the nanostructured poly(titanium oxide) uniformly distributed over the volume of the polymer matrix were synthesized from the titanium alkoxides using the sol-gel method in 2-hydroxyethyl methacrylate (HEMA) medium [33–35]. The materials showed a photochromic transition $Ti^{4+} \leftrightarrow Ti^{3+}$ with a quantum yield >50%. In [36], a new polymeric material was developed, containing nanostructured poly(titanium oxide) in an organic copolymers of hydroxyethyl methacrylate and acrylonitrile and modified by F-content agent, which demonstrate photocatalytic activity in the decomposition reaction of organic pollutants and the self-cleaning effect. Recent studies have also shown the prospect of using such hybrid materials in the field of photonics, especially for 3D laser microstructure and optical information recording [37].

The sol-gel method is commonly used for obtaining the organic-inorganic composites with a uniform distribution of the inorganic nanodispersed phase even at the molecular level [38–48]. The important feature of this process is the ability to regulate the structure of the inorganic component in the organic matrix by controlling the conditions of the hydrolysis-condensation reactions.

It is also of interest to obtain organo-inorganic hybrid materials containing poly(titanium oxide) in which interpenetrating polymer networks (IPNs) act as an organic matrix, especially since research in this direction has not yet been carried out. Among the variety of composite materials, interpenetrating polymer networks attract particular attention due to the possibility of both modifying the properties of cross-linked polymers and obtaining new materials with a wide range of different properties [49,50]. Interpenetrating polymer networks are a combination of two or more polymers in networks where a partial interlacing on the molecular scale is present in the matrix. There is no covalent bonding and therefore the polymers cannot be separated unless the chemical bonds are broken [50]. This morphology of interpenetrating polymer networks can lead to synergistic properties of the initial components. However, for IPNs, as for most organic polymers, such disadvantages as low mechanical properties and thermal stability are inherent. To overcome these defects, these materials can be strengthened by the addition of an inorganic component. For these purposes, silica was used successfully [51,52].

When creating the new composite materials, it is important to study the characteristics that determine their phase structure, thermal modes of use, stability and mechanical strength, especially since the structure and properties of organic-inorganic nanocomposites are largely dependent on the content of the nanofiller and its distribution in the organic matrix.

This work aims to study the effect of poly(titanium oxide) obtained using the sol-gel method in HEMA medium by varying the content of inorganic component and a molar ratio of titanium isopropoxide to water on viscoelastic and thermophysical properties of OI IPNs.

2. Experimental

The following reagents were used in this study: Titanium (IV) isopropoxide ((Ti(OPr^i)$_4$), 97.0% purity, Sigma-Aldrich) was used without further purification. 2-Hydroxyethyl methacrylate (HEMA, 99.3% purity, Merck, Darmstadt, Germany) was used without further purification. Toluylene diisocyanate (TDI, Merck, Darmstadt, Germany), which is a mixture of 2.4- and 2.6-toluylene diisocyanate (ratio 80/20) was distilled before use. Poly(propylene) glycol (M_n = 1000 g/mol) (PPG, Sigma-Aldrich, St. Louis, MO, USA) was dried under vacuum at (80 ± 5) °C for 4 h before the use. Trimethylolpropane (TMP, 99% purity, Merck, Darmstadt, Germany) was dried under vacuum at 40 °C for 5 h before the use. 2.2′-Azobis(2-methylbutyronitrile) (AIBN, Sigma-Aldrich, St. Louis, MO, USA) was recrystallized from ethanol before use. The solution of 0.1 N hydrochloric acid (HCl) was prepared from a standardized solution. The distilled H_2O was used as the solvent.

Organic-inorganic interpenetrating polymer networks are obtained on the basis of cross-linked polyurethane, poly(hydroxyethyl methacrylate) and poly(titanium oxide). Synthesis of poly(titanium oxide) was carried out by the hydrolytic polycondensation of Ti(OPr^i)$_4$ in the presence of hydrochloric acid in the medium of the initial component of IPNs-HEMA. The molar ratio of Ti(OPr^i)$_4$/HEMA was 1/16; 1/12 and 1/8, which calculated as TiO_2 was 3.8; 5.1 and 7.4 wt% respectively. The hydrolysis of Ti(OPr^i)$_4$ was carried out at the rate of Ti(OPr^i)$_4$/H_2O = 1/1 and 1/2 mol. The reaction mixture was stirred vigorously for 3 h followed by the formation of poly(titanium oxide) for 48 h. Finally, transparent orange-colored liquid systems were obtained and evacuated at 40 °C with a residual pressure of 10–20 mmHg for removing the by-products of hydrolysis-condensation reactions; water and isopropyl alcohol.

The urethane component of OI IPNs was obtained in two stages. At the first stage macrodiisocyanate (MDI) was synthesized by the interaction of TDI and PPG with the ratio NCO/OH = 2/1. At the second stage, TMP as the cross-linking agent was added to MDI at the molar ratio MDI/TMP = 3/2. The reaction was carried out at 70 °C and with vigorous stirring for 15 min.

Furthermore, HEMA containing poly(titanium oxide) and AIBN initiator of radical polymerization with a concentration of 0.025 mol/L was added to the urethane component to form OI IPNs. The initial IPNs were obtained on the basis of PU and PHEMA without poly(titanium oxide) as described above. The ratio of PU/PHEMA components in the initial and OI IPNs was 30/70 wt%. For obtaining the films of IPNs/OI IPNs, the reaction mixture was poured into sealed forms, followed by polymerization at 60 °C (20 h) and 100 °C (3 h).

IR spectra of HEMA-containing poly(titanium oxide) were recorded on a Bruker Tensor-37 Fourier spectrometer (Bruker Optics, Ettlingen, Germany) in the frequency range 400–4000 cm^{-1} with a resolution of 4 cm^{-1} by applying the test material to KBr plates.

The viscoelastic properties of the initial IPNs and OI IPNs samples were studied by DMA (Dynamic Mechanical Analyzer Q 800, TA Instruments, New Castle, DE, USA). To measure the viscoelastic properties, polymer films with a size of 50 mm × 5 mm (0.2–0.5) mm were used. The measurements were performed in the tension mode at a frequency of 10 Hz, and the heating rate was 2.0 °C/min.

The values of glass transition temperature (T_g) were determined by the position of the maximum of the tangent of mechanical losses (tanδ). The molecular weight of the chain segments between crosslinks (M_c) was calculated via the equation [53]:

$$M_c = \frac{3\rho \cdot R \cdot T}{E_\infty} \quad (1)$$

where ρ is the density of the polymer; R is the gas constant; T is the value of absolute temperature; and E_∞ is the value of the equilibrium elastic modulus.

Thermophysical properties of the initial IPNs and OI IPNs were studied by DSC method using Differential Scanning Calorimeter Q 2000 (TA Instruments, New Castle, DE, USA) in a nitrogen atmosphere in the temperature range from 273 to 693 K and heating rate of 2.0 °C/min.

Samples weighing 0.01–0.015 g were placed in aluminum capsules which were then sealed. The heating/cooling scan-mode was used. The midpoint of the endothermic transition on the curve of the temperature dependence of heat capacity (C_p) corresponded to T_g value of the polymer.

The fraction of the interfacial region ($1 - F$) was calculated according to the simplified Fried approximation for partially compatible multicomponent polymer systems [54] based on DSC data:

$$(1 - F) = 1 - \frac{W_1 \Delta C_{p1} + W_2 \Delta C_{p2}}{W_1 \Delta C_{p1}^0 + W_2 \Delta C_{p2}^0} \quad (2)$$

where: W_1 and W_2 are the mass fractions of the components in the mixture; ΔC_{p1} and ΔC_{p2} are the heat capacity increments of the phases that have released; and ΔC_{p1}^0 and ΔC_{p2}^0 are increments of heat capacity of individual polymers.

To analyze the morphology features of the investigated compositions, the film samples were split using liquefied nitrogen. The surface of the transverse cleavages was examined by scanning electron microscopy (SEM) using a JSM-35-C microscope (JEOL, Tokyo, Japan) at an accelerating voltage of 30 kV and an increase of x2000. To prevent the accumulation of surface charge and increase the contrast on the fractured surface a plasma deposition of a gold layer ~4 nm thick was carried out using a Neo Coater installation (JEOL, Tokyo, Japan).

3. Results and Discussion

To understand the processes occurring at the stage of the formation of poly(titanium oxide) in HEMA medium with varying the molar ratio Ti(OPr^i)$_4$/H$_2$O, the IR spectra were studied (Figure 1). The appearance of a low-intensity absorption band at 624 cm^{-1} for curves 2 and 3 relative to 1 is associated with the vibrations of Ti-O-Ti-groups of poly(titanium oxide) [55,56]. An increase in the relative intensity of ν C=O groups at 1720 cm^{-1} is observed 1.5 times for curve 2 and 1.2 times for curve 3 relative to the spectrum of the initial HEMA (curve 1), and the appearance of a new band at 1522 cm^{-1} is explained by the formation of complex interactions by the donor-acceptor mechanism (Figure 2) [57,58]. The relative intensity of the stretching band at 1170 cm^{-1} (ν C–O–) for poly(titanium oxide) in HEMA also increases by 1.5 and 1.2 times obtained at a molar ratio of Ti(OPr^i)$_4$/H$_2$O = 1/1 and Ti(OPr^i)$_4$/H$_2$O = 1/2, respectively. The increase in the relative intensity of the band at 1170 cm^{-1} is associated with the partial formation of poly(titanium oxide)-oxoethyl methacrylate (Figure 3) [59].

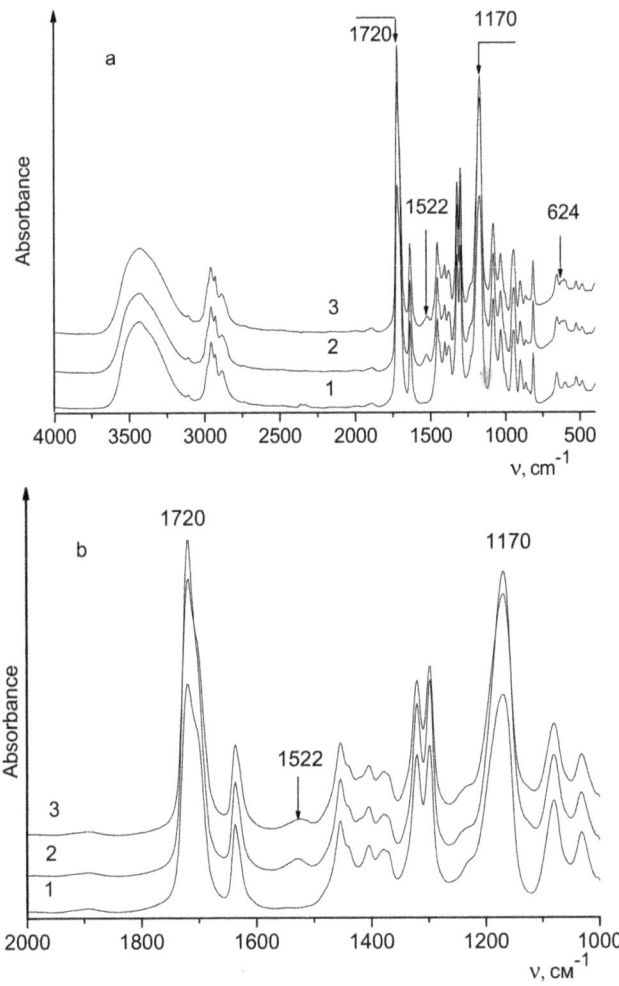

Figure 1. FTIR spectra (**a**,**b**) of HEMA (1) and poly(titanium oxide) in HEMA at molar ratio Ti(OPr^i)$_4$ to H$_2$O: 1/1 (2) and 1/2 (3).

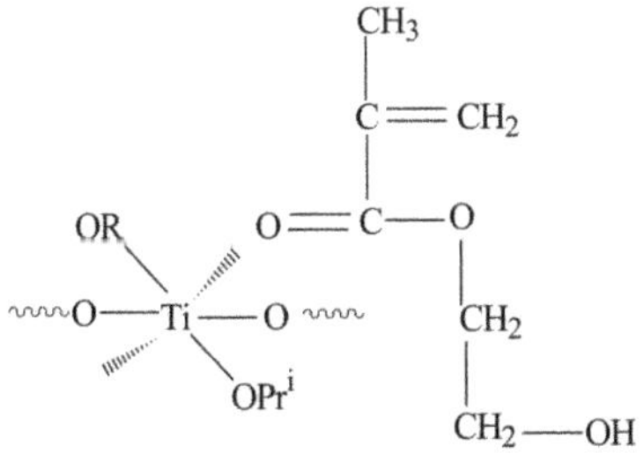

Figure 2. The complex interaction, where OPr^i is the alkoxide group.

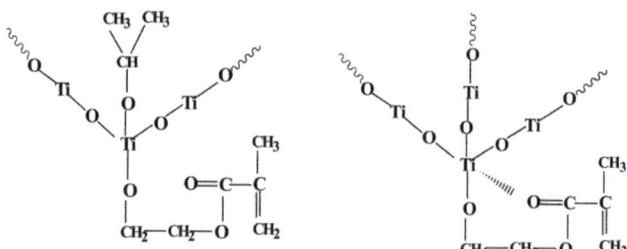

Figure 3. Possible structures of poly(titanium oxide) in HEMA obtained by varying the molar ratio Ti(OPr^i)$_4$/H$_2$O.

Thus, not only was the formation of poly(titanium oxide) in the HEMA medium during sol-gel synthesis established by using FTIR spectroscopy, but also the appearance of complex interactions by the donor-acceptor mechanism and the nucleophilic substitution reaction with the formation of poly(titanium oxide)-oxoethyl methacrylate. Additionally, it is worth noting that the amount of water that takes part in the hydrolysis of Ti(OPr^i)$_4$ directly affects both the presence of unreacted alkoxide groups and the structure of the resulting poly(titanium oxide) as a whole. In the middle of the twentieth century, R. Field and P. Kouv found that linear structure of poly(titanium oxide) is predominantly formed when Ti(OPr^i)$_4$/H$_2$O $\geq$ 1 ratio and branched one with a molar ratio of Ti(OPr^i)$_4$/H$_2$O < 1 (Figure 3) [60].

Dynamic mechanical analysis is one of the most widespread methods of studying the relaxation properties of polymer composites which allows the number of values (dynamic modulus, the tangent of mechanical losses, glass transition temperature, etc.) which have both fundamental and practical significance to be determined [53]. Therefore in this study, it was advisable to consider the influence of poly(titanium oxide) as well as its topology on the viscoelastic properties of OI IPNs. The temperature dependences of tanδ for IPNs and OI IPNs with the varying of the content of poly(titanium oxide) and the molar ratio of Ti(OPr^i)$_4$/H$_2$O are shown in Figure 4. The parameters of the relaxation transitions, the values of E_∞ and M_c for the studied systems, are presented in Table 1.

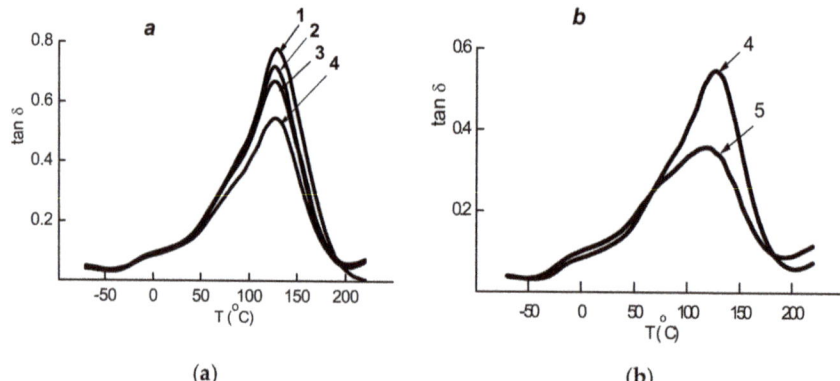

Figure 4. Temperature dependencies of tanδ (a,b) for the samples of the IPNs (1), OI IPNs-1 (2), OI IPNs-2 (3), OI IPNs-3 (4), OI IPNs-4 (5). The composition of the samples is given in Table 1.

Table 1. Viscoelastic properties of IPNs and OI IPNs.

Samples	PU/PHEMA/TiO$_2$ *, wt%	Ti(OPr^i)$_4$/H$_2$O, mol	PHEMA Phase		E_∞ MPa	M_c
			T_{g2}, °C	tanδ_{max2}		
IPNs	30.0/70.0/0	—	128	0.78	3.7	3760
OI IPNs-1	29.30/68.26/2.44	1/2	126	0.72	5.4	2600
OI IPNs-2	29.05/67.73/3.22	1/2	126	0.67	6.6	2130
OI IPNs-3	28.60/66.75/4.65	1/2	126	0.54	13.2	1070
OI IPNs-4	28.60/66.75/4.65	1/1	118	0.36	25.6	550

* the content of poly(titanium oxide) in terms of TiO$_2$; E_∞, the value of the equilibrium elastic modulus; M_c, the molecular weight of the chain segments between crosslinks.

The T_g at 128 °C corresponds to PHEMA [33] and the Tg at −5 °C corresponds to the PU polymer Figure 4a. Such temperature dependence tanδ indicates that IPNs is a two-phase system. Similar dependences of tanδ (T) are also observed for all samples (curves 2–5), which confirms their two-phase structure.

The study of the viscoelastic characteristics of OI IPNs showed that with an increase in poly(titanium oxide) content, there is a significant decrease in the relaxation transition height for the PHEMA phase (Figure 4a, curves 1–4, Table 1). The reduction in the intensity of the relaxation maximum with the increasing content of the inorganic component is often observed for the nanostructured hybrid systems. It is usually connected with the formation of the dense inorganic network that prevents the segmental mobility of the polymer chains [57,61]. In our case, the decrease in tanδ_{max2} is a result of blocking of the mobility of the significant fraction of the polymer segments (relaxators) involved in the relaxation transition for PHEMA phase. It may occur due to the partial formation of poly(titanium oxide)-oxoethyl methacrylate as a result of exchange reactions between non-hydrolyzed groups of poly(titanium oxide) and HEMA hydroxyl groups and also due to the formation of the complex interaction between Ti atoms and C=O-group of HEMA. The presence of such interactions is also supported by the fact that even a slight increase in poly(titanium oxide) content leads to a significant decrease in the M_c (the molecular weight of the chain segments between crosslinks) value in OI IPNs (Table 1). This indicates a significant increase in the number of cross-links and/or topological entanglements in the organic-inorganic hybrid polymer system.

It is an interesting fact that the presence of the inorganic component practically does not change the value of T_{g2} (Table 1). Although in the classical sense, the increase in the cross-linking density is usually accompanied by the increase in the glass transition temperature and a corresponding decrease in the M_c value. It can be assumed that poly(titanium oxide) in some way acts as a compatibilizer due to both the grafting of the inorganic component to the PHEMA macrochains, but also accidentally interacting with the urethane component due to highly active non-hydrolyzed alkoxide groups. According to the authors, this effect leads to the absence of an increase in the temperature of the acrylate component.

The comparison of the dependencies of tanδ(T) for OI IPNs-3 and OI IPNs-4 samples (Figure 4b) with the same content of poly(titanium oxide), but obtained at different molar ratios of Ti(OPr^i)$_4$/H$_2$O showed that the topology of inorganic component influences the phase structure of OI IPNs. This is evidenced by a change in the parameters of the relaxation transition for PHEMA phase, as well as a 2-fold decrease in the M_c value for OI IPNs-4 compared to OI IPNs-3 sample (Table 1). Perhaps, the linear structure of poly(titanium oxide) which was formed at the molar ratio of Ti(OPr^i)$_4$/H$_2$O = 1/1, promotes an increase in the number of the grafting and donor-acceptor interactions. It increases the number of polymer segments with locked mobility and reduces the height of tanδ of PHEMA phase (Figure 4b). However, at the branched structure of poly(titanium oxide) which was obtained at the molar ratio Ti(OPr^i)$_4$/H$_2$O = 1/2, the number of such interactions may be limited due to the occurrence of steric hindrances and the deficiency of alkoxide groups. The comparison of the M_c values for OI IPNs-3 and OI IPNs-4 samples (Table 1) shows that the only change in the topology of poly(titanium oxide) structure from the branched to the linear leads to the significant increase in the cross-linking density in OI IPNs.

The results obtained by DSC method also confirm that both the initial IPNs and OI IPNs have a two-phase structure as evidenced by the presence of two heat capacity jumps (ΔC_p) on the temperature dependence of $C_p = f(T)$. The glass transition temperature of PU (T_{g1}) and PHEMA (T_{g2}) were determined from $C_p = f(T)$ dependence (Table 2, Figure 5).

Table 2. Thermophysical parameters of relaxation transitions of IPNs and OI IPNs.

Samples	PU/PHEMA/ TiO_2 *, wt%	$Ti(OPr^i)_4/H_2O$, mol	T_{g1} PU-Enriched Phase, °C	T_{g2} PHEMA-Enriched Phase, °C	ΔC_{pPU} kJ/(kg K)	ΔC_{pPHEMA} kJ/(kg K)	$1 - F$
PU	100.0/0/0	-	−18.86	-	0.4936	-	-
PHEMA	0/100.0/0	-	-	64.51	-	0.21	-
IPNs	30.0/70.0/0	-	−18.44	53.32	0.1833	0.2802	0.24
OI IPNs-1	29.30/68.26/2.44	1/2	−13.78	53.19	0.2139	0.1886	0.41
OI IPNs-2	29.05/67.73/3.22	1/2	−14.33	53.53	0.1966	0.2248	0.34
OI IPNs-3	28.60/66.75/4.65	1/2	−14.65	53.08	0.231	0.1925	0.38

*the content of poly(titanium oxide) in terms of TiO_2; ΔC_{pPU}, change in the values of the heat capacity jump of the PU phase; ΔC_{pPHEMA}, change in the values of the heat capacity jump of the PHEMA phase.

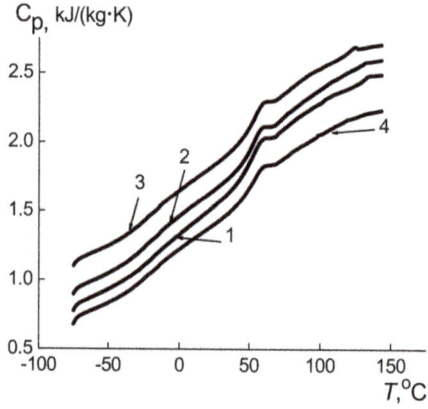

Figure 5. Temperature dependences of heat capacity for IPNs (1); OI IPNs-1 (2), OI IPNs-2 (3), OI IPNs-3 (4). The composition of the samples is given in Table 2.

The results presented in Table 2 show that the introduction of poly(titanium oxide) in IPNs slightly increases the glass transition temperature of the PU phase, but the glass transition temperature of the PEMA phase is practically unchanged. The value of the heat capacity increment for the PU phase slightly increases relative to the initial IPNs with the increase in poly(titanium oxide) content, and it decreases for the PHEMA phase. It should be noted that the nonlinear nature of the change in the values of the heat capacity jump in the PU and PHEMA phases in OI IPNs samples with increasing Ti component. It can be assumed that poly(titanium oxide) not only increases the rigidity of the PHEMA phase due to the partial formation of poly(titanium oxide)-oxoethyl methacrylate, but also possibly acts as a compatibilizer which leads to the absence of increasing T_{g2} value of PHEMA component and a slight increase this parameter for PU component. The results obtained by DSC method confirm the data obtained by DMA method.

Micro-phase separation processes also occur with the formation of the interfacial region (IFR) between the components of IPNs. Each phase can be considered quasi-equilibrium with the molecular level of mixing, but in general, these are systems where there is no deep penetration at the molecular level throughout the volume of the system.

The peculiarity of such systems is that for both IPNs and semi-interpenetrating polymer networks (semi-IPNs) IFR can be considered as a quasi-equal third phase. Its appearance is the result of a spinodal separation mechanism. Phase separation by the spinodal mechanism begins with the formation of a continuously interconnected periodic structure, which gradually shifts to droplet as a result of the decomposition of spinodal structures at the last stages of phase separation due to an increase in the interfacial tension [62,63].

The number of equations was proposed by various authors to evaluate IFR [53,64]. They were deduced on the assumption of additivity of heat capacity jumps in partially compatible systems. We used the simplified Fried approximation to calculate the share of IFR $((1 - F))$ (2).

Thus, IFR is absent if $(1 - F) = 0$; all polymers are in IFR if $(1 - F) = 1$. The value of the share of the interfacial region calculated in such way is relative.

As is shown in Table 2, the proportion of the interfacial region increases in OI IPNs compared to initial IPNs with increasing of poly(titanium oxide) obtained at a molar ratio of $Ti(OPr^i)_4/H_2O = 1/2$, which is probably related to the growth of compatibility of IPNs components. This effect is probably due to the partial grafting of poly(titanium oxide) to both PHEMA and PU macrochains during their formation, and/or with a decrease in the phase separation rate at the initial stages of composite formation due to the catalytic effect of the titanium component on the urethane formation reaction and increasing the viscosity of the reaction medium as a whole.

The scanning electron microscopy was used for a more complete understanding of the structure and morphology of the obtained IPNs and OI IPNs containing poly(titanium oxide) synthesized in a HEMA medium.

SEM studies have shown sufficient differences in the morphology of initial IPNs and OI IPNs. The presence of characteristic morphology of the fractured piece of IPNs film, as well as the distribution of color shades in the obtained images (Figure 6a) may indicate phase separation of the components (PU and PHEMA) in the composite.

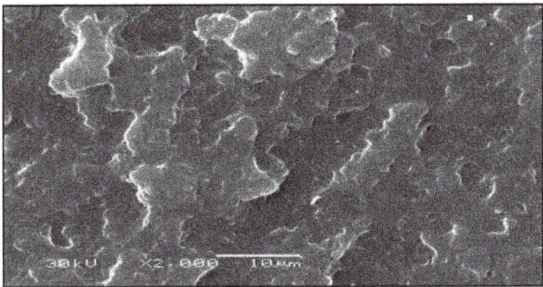

(a) PU/PHEMA 30.0/70.0, wt%

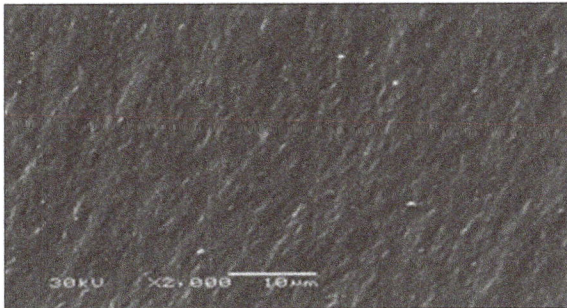

(b) PU/PHEMA/TiO₂ 28.60/66.75/4.65, wt%

Figure 6. SEM images of transverse cleavages of IPNs (**a**) and OI IPNs (**b**) samples.

However, the significant smoothing of the relief of the transverse cleavage surface of OI IPNs sample and the decrease in the color contrast level (Figure 6b) confirm that the introduction of poly(titanium oxide) into the polymer matrix significantly reduces the phase separation level of the hybrid composition.

The obtained results indirectly confirm the presence of the compatibilizing effect of poly(titanium oxide), as well as its influence on the relaxation and thermophysical behavior of the studied samples. Additionally, it is worth noting that the distribution of the inorganic component in the polymer matrix is quite uniform without significant aggregation, which may indicate the presence of the appropriate physicochemical interaction between them. This is especially important for the manufacture of appropriate fluorescent matrices based on TiO_2, when the latter is doped with a luminescent impurity [65–67].

4. Conclusions

It was established by using DMA and DSC methods that the OI IPNs system is two-phase; however, the increase in the interfacial region share with the increase in poly(titanium oxide) content as well as their morphology according to SEM indicates that the inorganic component acts as a compatibilizer. It was shown by using the DMA method that the increase in poly(titanium oxide) content in OI IPNs leads to a decrease in the relaxation maximum intensity of PHEMA component and the reduction in the molecular weight of the chain segments between the cross-links. Such changes indicate an increase in the effective cross-linking density of OI IPNs due to the partial grafting of poly(titanium oxide) to PEMA. The decrease in the segmental mobility of the polymer chains due to the formation of poly(titanium oxide)-oxoethyl methacrylate is indicated by a decrease in the heat capacity increment of PHEMA component of OI IPNs relative to initial IPNs. It has been shown that the topology of poly(titanium oxide) structure significantly influences the relaxation behavior of the samples.

Author Contributions: Conceptualization, T.T. and A.I.P.; methodology, T.T.; software, T.T.; formal analysis, T.T. and A.I.P.; investigation, T.T.; resources, A.I.P.; data curation, A.I.P.; writing—original draft preparation, T.T. and A.I.P.; writing—review and editing, T.T. and A.I.P.; visualization, T.T.; supervision, A.I.P.; project administration, A.I.P.; funding acquisition, A.I.P. All authors have read and agreed to the published version of the manuscript.

Funding: This study was partly supported by the M-ERA.NET project SunToChem.

Institutional Review Board Statement: Not applicable.

Informed Consent Statement: Not applicable.

Data Availability Statement: Not applicable.

Acknowledgments: The authors thank V. Serga for many useful discussions. The research was (partly) performed in the Institute of Solid State Physics, University of Latvia ISSP UL. ISSP UL as the Center of Excellence is supported through the Framework Program for European universities Union Horizon 2020, H2020-WIDESPREAD-01–2016–2017-Teaming Phase 2 under Grant Agreement No. 739508, CAMART2 project.

Conflicts of Interest: The authors declare no conflict of interest.

References

1. Pandey, S.; Mishra, S.B. Sol–gel derived organic–inorganic hybrid materials: Synthesis, characterizations and applications. *J. Sol Gel Sci. Technol.* **2011**, *59*, 73–94. [CrossRef]
2. Canto, C.F.; Prado, L.D.A.; Radovanovic, E.; Yoshida, I.V.P. Organic–inorganic hybrid materials derived from epoxy resin and polysiloxanes: Synthesis and characterization. *Polym. Eng. Sci.* **2008**, *48*, 141–148. [CrossRef]
3. Wang, S.; Kang, Y.; Wang, L.; Zhang, H.; Wang, Y.; Wang, Y. Organic/inorganic hybrid sensors: A review. *Sens. Actuators Chem. B* **2013**, *182*, 467–481. [CrossRef]
4. Karbovnyk, I.; Klym, H.; Piskunov, S.; Popov, A.A.; Chalyy, D.; Zhydenko, I.; Lukashevych, D. The impact of temperature on electrical properties of polymer-based nanocomposites. *Low Temp. Phys.* **2020**, *46*, 1231–1234. [CrossRef]

5. Aksimentyeva, O.I.; Savchyn, V.P.; Dyakonov, V.P.; Piechota, S.; Horbenko, Y.Y.; Opainych, I.Y.; Demchenko, P.Y.; Popov, A.; Szymczak, H. Modification of polymer-magnetic nanoparticles by luminescent and conducting substances. *Mol. Cryst. Liq. Cryst.* **2014**, *590*, 35–42. [CrossRef]
6. Savchyn, V.P.; Popov, A.I.; Aksimentyeva, O.I.; Klym, H.; Horbenko, Y.Y.; Serga, V.; Moskina, A.; Karbovnyk, I. Cathodoluminescence characterization of polystyrene-$BaZrO_3$ hybrid composites. *Low Temp. Phys.* **2016**, *42*, 597–600. [CrossRef]
7. Karbovnyk, I.; Olenych, I.; Kukhta, A.; Lugovskii, V.; Sasnouski, G.; Olenych, Y.; Luchechko, A.; Popov, A.I.; Yarytska, L. Multicolor photon emission from organic thin films on different substrates. *Radiat. Meas.* **2016**, *90*, 38–42. [CrossRef]
8. Zhyhailo, M.; Yevchuk, I.; Yatsyshyn, M.; Korniy, S.; Demchyna, O.; Musiy, R.; Raudonis, R.; Zarkov, A.; Kareiva, A. Preparation of polyacrylate/silica membranes for fuel cell application by in situ UV polymerization. *Chemija* **2020**, *31*, 247–254. [CrossRef]
9. Laurikėnas, A.; Mažeika, K.; Baltrūnas, D.; Skaudžius, R.; Beganskienė, A.; Kareiva, A. Hybrid organic-inorganic Fe_3O(TFBDC)$_3$(H$_2$O)$_3$·(DMF)$_3$ compound synthesized by slow evaporation method: Characterization and comparison of magnetic properties. *Lith. J. Phys.* **2020**, *60*, 78. [CrossRef]
10. Kiele, E.; Lukseniene, J.; Grigucevicienė, A.; Selskis, A.; Senvaitiene, J.; Ramanauskas, R.; Raudonis, R.; Kareiva, A. Methyl–modified hybrid organic-inorganic coatings for the conservation of copper. *J. Cult. Herit.* **2014**, *15*, 242–249. [CrossRef]
11. Aksimentyeva, O.; Konopelnyk, O.; Bolesta, I.; Karbovnyk, I.; Poliovyi, D.; Popov, A.I. Charge transport in electrically responsive polymer layers. *J. Phys. Conf. Ser.* **2007**, *93*, 012042. [CrossRef]
12. Aksimentyeva, O, I.; Chepikov, I.B.; Filipsonov, R.V.; Malynych, S.Z.; Gamernyk, R.V.; Martyniuk, G.V.; Horbenko, Y.Y. Hybrid composites with low reflection of IR radiation. *Phys. Chem. Solid State* **2020**, *21*, 764–770. [CrossRef]
13. Kickelbick, G. Introduction to hybrid materials. In *Hybrid Materials: Synthesis, Characterization, and Applications*; Wiley-VCH Verlag GmbH & Co.: Weinheim, Germany, 2007; pp. 1–48.
14. Sabah, F.A.; Razak, I.A.; Kabaa, E.A.; Zaini, M.F.; Omar, A.F. Characterization of hybrid organic/inorganic semiconductor materials for potential light emitting applications. *Opt. Mater.* **2020**, *107*, 110117. [CrossRef]
15. Quy, H.V.; Truyen, D.H.; Kim, S.; Bark, C.W. Facile Synthesis of Spherical TiO_2 Hollow Nanospheres with a Diameter of 150 nm for High-Performance Mesoporous Perovskite Solar Cells. *Materials* **2021**, *14*, 629. [CrossRef]
16. Li, Y.; Wang, W.; Wang, F.; Di, L.; Yang, S.; Zhu, S.; Yao, Y.; Ma, C.; Dai, B.; Yu, F. Enhanced Photocatalytic Degradation of Organic Dyes via Defect-Rich TiO_2 Prepared by Dielectric Barrier Discharge Plasma. *Nanomaterials* **2019**, *9*, 720. [CrossRef]
17. Pant, B.; Park, M.; Park, S.-J. Recent Advances in TiO_2 Films Prepared by Sol-Gel Methods for Photocatalytic Degradation of Organic Pollutants and Antibacterial Activities. *Coatings* **2019**, *9*, 613. [CrossRef]
18. Ramirez, L.; Ramseier Gentile, S.; Zimmermann, S.; Stoll, S. Behavior of TiO_2 and CeO_2 Nanoparticles and Polystyrene Nanoplastics in Bottled Mineral, Drinking and Lake Geneva Waters. Impact of Water Hardness and Natural Organic Matter on Nanoparticle Surface Properties and Aggregation. *Water* **2019**, *11*, 721. [CrossRef]
19. Zhao, W.; Yang, X.; Liu, C.; Qian, X.; Wen, Y.; Yang, Q.; Sun, T.; Chang, W.; Liu, X.; Chen, Z. Facile Construction of All-Solid-State Z-Scheme g-C_3N_4/TiO_2 Thin Film for the Efficient Visible-Light Degradation of Organic Pollutant. *Nanomaterials* **2020**, *10*, 600. [CrossRef] [PubMed]
20. Regmi, C.; Lotfi, S.; Espíndola, J.C.; Fischer, K.; Schulze, A.; Schäfer, A.I. Comparison of Photocatalytic Membrane Reactor Types for the Degradation of an Organic Molecule by TiO_2-Coated PES Membrane. *Catalysts* **2020**, *10*, 725. [CrossRef]
21. Dukenbayev, K.; Kozlovskiy, A.; Kenzhina, I.; Berguzinov, A.; Zdorovets, M. Study of the effect of irradiation with Fe^{7+} ions on the structural properties of thin TiO_2 foils. *Mater. Res. Express* **2019**, *6*, 046309. [CrossRef]
22. Knoks, A.; Kleperis, J.; Grinberga, L. Raman spectral identification of phase distribution in anodic titanium dioxide coating. *Proc. Estonian Acad. Sci.* **2017**, *66*, 422–429. [CrossRef]
23. Kozlovskiy, A.; Shlimas, D.; Kenzhina, I.; Boretskiy, O.; Zdorovets, M. Study of the Effect of Low-Energy Irradiation with O^{2+} Ions on Radiation Hardening and Modification of the Properties of Thin TiO_2 Films. *J. Inorg. Organomet. Polym. Mater.* **2021**, *31*, 790–801. [CrossRef]
24. Mattsson, M.S.M.; Azens, A.; Niklasson, G.A.; Granqvist, C.G.; Purans, J. Li intercalation in transparent Ti–Ce oxide films. Energetics and ion dynamics. *J. Appl. Phys.* **1997**, *81*, 6432–6437. [CrossRef]
25. Prashantha, K.; Rashmi, B.J.; Venkatesha, T.V.; Lee, J.-H. Spectral characterization of apatite formation on poly(2-hydroxyethylmethacrylate)–TiO_2 nanocomposite film prepared by sol–gel process. *Spectrochim. Acta Part A Mol. Biomol. Spectrosc.* **2006**, *65*, 340–344. [CrossRef] [PubMed]
26. Štengl, V.; Houšková, V.; Bakardjieva, S.; Murafa, N.; Havlín, V. Optically Transparent Titanium Dioxide Particles Incorporated in Poly(hydroxyethyl methacrylate) Thin Layers. *J. Phys. Chem. C* **2008**, *112*, 19979–19985. [CrossRef]
27. Toledo, L.; Racine, E.; Pérez, V.; Henríquez, J.P.; Auzely-Velty, R.; Urbano, B.F. Physical nanocomposite hydrogels filled with low concentrations of TiO_2 nanoparticles: Swelling, networks parameters and cell retention studies. *Mater. Sci. Eng. C* **2018**, *92*, 769–778. [CrossRef]
28. Mubarak, S.; Dhamodharan, D.; Divakaran, N.; Kale, M.B.; Senthil, T.; Wu, L.; Wang, J. Enhanced Mechanical and Thermal Properties of Stereolithography 3D Printed Structures by the Effects of Incorporated Controllably Annealed Anatase TiO_2 Nanoparticles. *Nanomaterials* **2020**, *10*, 79. [CrossRef]
29. Moustafa, H.; Darwish, D.; Youssef, A.S.; El-Wakil, R.A. High-Performance of Nanoparticles and Their Effects on the Mechanical, Thermal Stability and UV-Shielding Properties of PMMA Nanocomposites. *Egypt. J. Chem.* **2018**, *61*, 23–32. [CrossRef]

30. Motaung, T.E.L.; Luyt, A.S.; Bondioli, F.; Messori, M.; Saladino, M.L.; Spinella Al Nasillo, G.; Caponetti, E. PMMA–titania nanocomposites: Properties and thermal degradation behavior. *Polym. Degrad. Stab.* **2012**, *97*, 1325–1333. [CrossRef]
31. Gayvoronsky, V.; Galas, A.; Shepelyavyy, E.; Dittrich, T.; Timoshenko, V.; Nepijko, S.; Brodyn, M.; Koch, F. Giant nonlinear optical response of nanoporous anatase layers. *Appl. Phys. B* **2005**, *80*, 97–100. [CrossRef]
32. Kuznetsov, A.I.; Kameneva, O.; Alexandrov, A.; Bityurin, N.; Marteau, P.; Chhor, K.; Sanchez, C.; Kanaev, A. Light-induced charge separation and storage in titanium oxide gels. *Phys. Rev. E* **2005**, *71*, 021403. [CrossRef]
33. Salomatina, E.V.; Bityurin, N.M.; Gulenova, M.V.; Gracheva, T.A.; Drozdov, M.N.; Knyazev, A.V.; Kir'yanov, K.V.; Markin, A.V.; Smirnova, L.A. Synthesis, structure, and properties of organic-inorganic nanocomposites containing poly(titanium oxide). *J. Mater. Chem. C* **2013**, *1*, 6375–6385. [CrossRef]
34. Salomatina, E.V.; Moskvichev, A.N.; Knyazev, A.V.; Smirnova, L.A. Effect of kinetic features in synthesis of hybrid copolymers based on Ti(O^{i}Pr)$_4$ and hydroxyethyl methacrylate on their structure and properties. *Russian J. Appl. Chem.* **2015**, *88*, 197–207. [CrossRef]
35. Kameneva, O.; Kuznestov, A.I.; Smirnova, L.A.; Rozes, L.; Sanchez, C.; Alexandrov, A.; Bityurin, N.; Chhor, K.; Kanaev, A. New photoactive hybrid organic-inorganic materials based on titanium-oxo-PHEMA nanocomposites exhibiting mixed valence properties. *J. Mater. Chem.* **2005**, *15*, 3380–3383. [CrossRef]
36. Ryabkova, O.; Redina, L.; Salomatina, E.; Smirnova, L. Hydrophobizated poly(titanium oxide) containing polymeric surfaces with UV-induced reversible wettability and self-cleaning properties. *Surf. Interfaces* **2020**, *18*, 100452. [CrossRef]
37. Fadeeva, E.; Koch, J.; Chichkov, B.; Kuznetsov, A.; Kameneva, O.; Bityurin, N.; Sanchez, C.; Kanaev, A. Laser imprinting of 3D structures in gel-based titanium oxide organic-inorganic hybrids. *Appl. Phys. A* **2006**, *84*, 27–30. [CrossRef]
38. Brinker, C.J.; Scherer, G.W. *Sol-Gel Science*; Academic Press: New York, NY, USA, 1990; 881p.
39. Jonauske, V.; Stanionyte, S.; Chen, S.-W.; Zarkov, A.; Juskenas, R.; Selskis, A.; Matijosius, T.; Yang, T.C.K.; Ishikawa, K.; Ramanauskas, R.; et al. Characterization of Sol-Gel Derived Calcium Hydroxyapatite Coatings Fabricated on Patterned Rough Stainless Steel Surface. *Coatings* **2019**, *9*, 334. [CrossRef]
40. Mura, S.; Ludmerczki, R.; Stagi, L.; Garroni, S.; Carbonaro, C.M.; Ricci, P.C.; Casula, M.F.; Malfatti, L.; Innocenzi, P. Integrating sol-gel and carbon dots chemistry for the fabrication of fluorescent hybrid organic-inorganic films. *Sci. Rep.* **2020**, *10*, 4770. [CrossRef] [PubMed]
41. Valeikiene, L.; Roshchina, M.; Grigoraviciute-Puroniene, I.; Prozorovich, V.; Zarkov, A.; Ivanets, A.; Kareiva, A. On the Reconstruction Peculiarities of Sol–Gel Derived Mg$_{2-x}$M$_x$/Al$_1$ (M = Ca, Sr, Ba) Layered Double Hydroxides. *Crystals* **2020**, *10*, 470. [CrossRef]
42. Tsvetkov, N.; Larina, L.; Ku Kang, J.; Shevaleevskiy, O. Sol-Gel Processed TiO$_2$ Nanotube Photoelectrodes for Dye-Sensitized Solar Cells with Enhanced Photovoltaic Performance. *Nanomaterials* **2020**, *10*, 296. [CrossRef]
43. Veselov, G.B.; Karnaukhov, T.M.; Bauman, Y.I.; Mishakov, I.V.; Vedyagin, A.A. Sol-Gel-Prepared Ni-Mo-Mg-O System for Catalytic Transformation of Chlorinated Organic Wastes into Nanostructured Carbon. *Materials* **2020**, *13*, 4404. [CrossRef] [PubMed]
44. Korolkov, I.V.; Kuandykova, A.; Yeszhanov, A.B.; Güven, O.; Gorin, Y.G.; Zdorovets, M.V. Modification of PET Ion-Track Membranes by Silica Nanoparticles for Direct Contact Membrane Distillation of Salt Solutions. *Membranes* **2020**, *10*, 322. [CrossRef]
45. Smalenskaite, A.; Pavasaryte, L.; Yang, T.C.K.; Kareiva, A. Undoped and Eu^{3+} Doped Magnesium-Aluminium Layered Double Hydroxides: Peculiarities of Intercalation of Organic Anions and Investigation of Luminescence Properties. *Materials* **2019**, *12*, 736. [CrossRef] [PubMed]
46. Jaafar, A.; Hecker, C.; Árki, P.; Joseph, Y. Sol-Gel Derived Hydroxyapatite Coatings for Titanium Implants: A Review. *Bioengineering* **2020**, *7*, 127. [CrossRef] [PubMed]
47. Grazenaite, E.; Garskaite, E.; Stankeviciute, Z.; Raudonyte-Svirbutaviciene, E.; Zarkov, A.; Kareiva, A. Ga-Substituted Cobalt-Chromium Spinels as Ceramic Pigments Produced by Sol–Gel Synthesis. *Crystals* **2020**, *10*, 1078. [CrossRef]
48. Karbovnyk, I.; Borshchyshyn, I.; Vakhula, Y.; Lutsyuk, I.; Klym, H.; Bolesta, I. Impedance characterization of Cr^{3+}, Y^{3+} and Zr^{4+} activated forsterite nanoceramics synthesized by sol–gel method. *Ceram. Int.* **2016**, *42*, 8501–8504. [CrossRef]
49. Sperling, L.H. *Interpenetrating Polymer Networks and Related Materials*; Springer Science & Business Media: Cham, Switzerland, 2012.
50. Lipatov, Y.S.; Alekseeva, T. Phase-separated interpenetrating polymer networks. In *Advances in Polymer Science*; Springer: Berlin, Germany, 2007; Volume 208, pp. 147–194. ISBN 978-3-540-73071-2.
51. Widmaier, J.-M.; Bonilla, G. In situ synthesis of optically transparent interpenetrating organic/inorganic networks. *Polym. Adv. Technol.* **2006**, *17*, 634–640. [CrossRef]
52. Bonilla, G.; Martinez, M.; Mendoza, A.M.; Widmaier, J.-M. Ternary interpenetrating networks of polyurethane-poly(methyl methacrylate)–silica. Preparation by sol-gel process and characterization of films. *Eur. Polym. J.* **2006**, *42*, 2977–2986. [CrossRef]
53. Nielsen, L.E.; Landel, R.F. *Mechanical Properties of Polymers and Composites*; CRC Press: Boca Raton, FL, USA, 1993; p. 580, ISBN 9780824789640.
54. Beckman, E.J.; Karasz, F.E.; Porter, R.S.; MacKnight, W.J.; Van Hunsel, J.; Koningsveld, R. Estimation of the interfacial fraction in partially miscible polymer blends from DSC measurements. *Macromolecules* **1988**, *21*, 1193–1194. [CrossRef]
55. Lee, L.H.; Chen, W.C. High-Refractive-index thin films prepared from trialkoxysilane-capped poly(methyl methacrylate)–titania materials. *Chem. Mater.* **2001**, *13*, 1137–1142. [CrossRef]

56. Bach, L.G.; Islam, M.R.; Seo, S.Y.; Lim, K.T. A novel route for the synthesis of poly (2-hydroxyethyl methacrylate) grafted TiO$_2$ nanoparticles via surface thiol-lactam initiated radical polymerization. *J. Appl. Polym. Sci.* **2013**, *127*, 261–269. [CrossRef]
57. Wu, C.S. In situ polymerization of titanium isopropoxide in polycaprolactone: Properties and characterization of the hybrid nanocomposites. *J. Appl. Polym. Sci.* **2004**, *92*, 1749–1757. [CrossRef]
58. Chen, W.; Lee, S.; Lee, L.; Lin, J. Synthesis and Characterization of Trialkoxysilane-Capped Poly (Methyl Methacrylate)-titania Hybrid Optical Thin Films. *J. Mater. Chem.* **1999**, *9*, 2999–3003. [CrossRef]
59. Rozenberg, B.A.; Boiko, G.N.; Bogdanova, L.M.; Dzhavadyan, E.A.; Komarov, B.A. Mechanism of anionic polymerization of 2-hydroxyethyl (meth) acrylates initiated by alkali metals and their alkoxides. *Polym. Sci. Ser. A* **2003**, *45*, 819–825.
60. Feld, R.; Cowe, P.L. *Organic Chemistry of Titanium*; Butterworth & Co Publishers Ltd.: London, UK, 1965; 222p, ISBN 13 978-0408283007.
61. Trabelsi, S.; Janke, A.; Hässler, R.; Zafeiropoulos, N.E.; Fornasieri, G.; Bocchini, S.; Rozes, L.; Stamm, M.; Gérard, J.-F.; Sanchez, C. Novel Organo-functional titanium−oxo-cluster-based hybrid materials with enhanced thermomechanical and thermal properties. *Macromolecules* **2005**, *38*, 6068–6078. [CrossRef]
62. Ignatova, T.D.; Kosyanchuk, L.F.; Todosiychuk, T.T.; Nesterov, A.E. Reaction-induced phase separation and structure formation in polymer blends. *Compos. Interfaces* **2011**, *18*, 185–236. [CrossRef]
63. Lipatov, Y.S.; Nesterov, A.E. *Thermodynamics of Polymer Blends*; Techn. Publ. Co.: Basel, Switzerland, 1997; 450p.
64. Hourston, D.J.; Schäfer, F.U. Polyurethane/Polystyrene Oneshot Interpenetrating Polymer Networks with Good Damping Ability: Transition Broadening Through Crosslinking, Internetwork Grafting and Compatibilization. *Polym. Adv. Technol.* **1996**, *7*, 273–280. Available online: https://onlinelibrary.wiley.com (accessed on 28 February 2021). [CrossRef]
65. Serga, V.; Burve, R.; Krumina, A.; Romanova, M.; Kotomin, E.A.; Popov, A.I. Extraction–Pyrolytic Method for TiO$_2$ Polymorphs Production. *Crystals* **2021**, *11*, 431. [CrossRef]
66. Serga, V.; Burve, R.; Krumina, A.; Pankratova, V.; Popov, A.I.; Pankratov, V. Study of phase composition, photocatalytic activity, and photoluminescence of TiO$_2$ with Eu additive produced by the extraction-pyrolytic method. *J. Mater. Res. Technol.* **2021**, *13*, 2350–2360. [CrossRef]
67. Serikov, T.M.; Ibrayev, N.K.; Isaikina, O.Y.; Savilov, S.V. Nanocrystalline TiO$_2$ Films: Synthesis and Low-Temperature Luminescent and Photovoltaic Properties. *Russ. J. Inorg. Chem.* **2021**, *66*, 117–123. [CrossRef]

MDPI
St. Alban-Anlage 66
4052 Basel
Switzerland
Tel. +41 61 683 77 34
Fax +41 61 302 89 18
www.mdpi.com

Crystals Editorial Office
E-mail: crystals@mdpi.com
www.mdpi.com/journal/crystals

www.ingramcontent.com/pod-product-compliance
Lightning Source LLC
LaVergne TN
LVHW070614100526
838202LV00012B/644